Intelligent Techniques for Predictive Data Analytics

IEEE Press
445 Hoes Lane
Piscataway, NJ 08854

Intelligent Techniques for Predictive Data Analytics

Edited by

Neha Singh
Manipal University Jaipur
Jaipur, Rajasthan
India

Shilpi Birla
Manipal University Jaipur
Jaipur, Rajasthan
India

Mohd Dilshad Ansari
SRM University
Delhi-NCR, Haryana
India

Neeraj Kumar Shukla
King Khalid University
Abha
Saudi Arabia

IEEE Press

WILEY

Published by John Wiley & Sons, Inc., Hoboken, New Jersey.
Published simultaneously in Canada.

Library of Congress Cataloging-in-Publication Data
Names: Singh, Neha (Electronic engineering professor), editor. | Birla, Shilpi, editor. | Ansari, Mohd Dilshad, editor. | Neeraj Kumar Shukla, editor. | John Wiley & Sons, publisher.
Title: Intelligent techniques for predictive data analytics / Neha Singh, Shilpi Birla, Mohd Dilshad Ansari, Neeraj Kumar Shukla.
Description: Hoboken, New Jersey : Wiley, [2024] | Includes index.
Identifiers: LCCN 2024005729 (print) | LCCN 2024005730 (ebook) | ISBN 9781394227969 (hardback) | ISBN 9781394227983 (adobe pdf) | ISBN 9781394227976 (epub)
Subjects: LCSH: Engineering–Data processing. | Predictive analytics–Industrial applications. | Predictive analytics–Scientific applications.
Classification: LCC TA340 .I56 2023 (print) | LCC TA340 (ebook) | DDC 620.00285–dc23/eng/20240401
LC record available at https://lccn.loc.gov/2024005729
LC ebook record available at https://lccn.loc.gov/2024005730

Cover Design: Wiley
Cover Image: © Blue Planet Studio/Shutterstock

Set in 9.5/12.5pt STIXTwoText by Straive, Pondicherry, India

Contents

About the Editors

Dr. Neha Singh is a well-rounded and dynamic teaching professional with 19+ years of in-depth experience in educating undergraduate and postgraduate engineering students in electronics and communication engineering. She is currently working as Assistant Professor (Selection Grade) in the Department of Electronics and Communication Engineering at Manipal University Jaipur, Rajasthan, India. She did her PhD in the year 2020. Her areas of research interest encompass image processing, machine learning, data analysis, and VLSI design. She holds one Indian patent and has several papers published in journals and conferences of repute as well as many book chapters in indexed books. She has co-authored engineering textbooks and serves as an editor for indexed books. She has served as a reviewer for various international and peer-reviewed journals and conferences. She has also convened various conferences, workshops, and other events, chaired sessions, and organized summer internships in diode fabrication. She has guided several M Tech dissertations and B Tech projects and is guiding PhD scholars as well. She is a senior member of the IEEE.

Dr. Shilpi Birla is currently serving as Associate Professor (Senior Scale) in the Electronics and Communication Department at Manipal University Jaipur with over 16 years of experience in both academia and industry. She did her PhD in low power VLSI design. With a specialization in low power VLSI design, memory circuits, digital VLSI circuits, nanodevices, and image processing, she has made significant contributions to her field. Her research endeavors are reflected in over 60 research papers published in journals of repute and at various international conferences. She has played a pivotal role in organizing workshops covering HSPICE, TCAD, and

XILINX, as well as overseeing summer internships in diode fabrication and faculty development programs. Her commitment to fostering academic growth extends to guiding M Tech students and mentoring PhD candidates. Beyond her academic pursuits, she has actively participated in the academic community, serving as a session chair, conference steering committee member, editorial board member, and reviewer for international and national IEEE journals and conferences. She is also a senior member of the IEEE.

Dr. Mohd Dilshad Ansari is currently working as an associate professor in the Department of Computer Science and Engineering at SRM University Delhi-NCR, Sonepat, Haryana, India. He obtained his PhD and M Tech in Computer Science and Engineering from Jaypee University of Information Technology, Waknaghat, Solan, HP, India in 2018 and 2011, respectively. He received B Tech in Information Technology from Uttar Pradesh Technical University, Lucknow, Uttar Pradesh, in 2009. He has more than 12 years of academic/research experience; He has published more than 80 papers in international journals (SCIE/Scopus) and conferences (IEEE/Springer). He is the member of various technical/professional societies such as IEEE, UACEE, and IACSIT. He has been appointed as Editorial/Reviewer Board and Technical Programme Committee member in numerous reputed journals/conferences. He is also serving as an associate, academic, and guest editor in reputed journals and organized special sessions in IEEE/Springer conferences. His research interests include digital and fuzzy image processing, machine learning, IoT, and cloud computing.

Dr. Neeraj Kumar Shukla is currently working as an Associate Professor in the Department of Electrical Engineering at King Khalid University, Abha, Kingdom of Saudi Arabia. He has academic, research, and industry experience of more than 20 years. He did his Ph.D. in Low Power VLSI Design. He has authored more than 110 research publications in Journals of repute and International Conferences, apart from this he has 2 Patents, 1 PhD completed under him. He guided 20 M. Tech Dissertations, 50 B. Tech Projects and conducted 50 Short terms skills development training Programs and 30 expert lectures delivered. He is reviewer in several international conferences and Journal. His research areas are Low Power VLSI Design, Digital Design, Machine Learning.

List of Contributors

Karan Aggarwal
Electronics and Communication Engineering Department
Maharishi Markandeshwar (Deemed to be University)
Ambala, India

Poojan Agrawal
Information and Communication Technology, Pandit Deendayal Energy University, Gandhinagar
Gujarat, India

Sayed Q.Y. Al Khalidi
Department of Information Technology, King Khalid University
Abha, Kingdom of Saudi Arabia

Mohd Dilshad Ansari
Department of Computer Science and Engineering
SRM University Delhi-NCR
Sonepat, India

SMD Azash
Department of Law
School of Law Presidency University
Bengaluru, Karnataka, India

Indu Bala
School of Electrical and Electronics Engineering, Lovely Professional University, Punjab, India

Paolo Barsocchi
Wireless Networks (WN)
Institute of Information Science and Technologies, National Research Council, Pisa, Italy

Pankaj Bhambri
Department of Information Technology, Guru Nanak Dev Engineering College, Ludhiana
Punjab, India

Akash Kumar Bhoi
Wireless Networks (WN)
Institute of Information Science and Technologies, National Research Council, Pisa, Italy; Directorate of Research, Sikkim Manipal University
Gangtok, Sikkim, India

Ruchi Doshi
Department of Computer Science and Engineering, Universidad Azteca
Chalco, Mexico

Nafees Akhter Farooqui
Department of CSE, Koneru Lakshmaiah Education Foundation
Vaddeswaram, Andhra Pradesh, India

Vijay Gali
Department of Electrical & Computer Engineering, Nazarbayev University
Astana, Kazakhstan

Deepika Ghai
School of Electronics and Electrical Engineering, Lovely Professional University, Phagwara, India

Umashankar Ghugar
Department of CSE
OP Jindal University
Raigarh, India

Murat Gök
Department of Computer Engineering
Yalova University, Yalova, Turkey

Nand Kishor Gupta
Department of Electical & Electronic Engineering, Poornima University
Jaipur, India

Mohammed Sayeeduddin Habeeb
Department of Electrical Engineering
College of Engineering, King Khalid University, Abha, KSA

Mohd. Haleem
Department of Computer Science
Era University, Lucknow
Uttar Pradesh, India

Kamal Kant Hiran
School of Computer Science and IT, Symbiosis University of Applied Sciences, Indore, India

Mohammad Ishrat
Department of CSE
Koneru Lakshmaiah Education Foundation, Vaddeswaram
Andhra Pradesh, India

Sardar M.N. Islam
Institute for Sustainable Industries and Liveable Cities, Victoria University
Melbourne, Australia

Wasim Khan
Department of CSE
Koneru Lakshmaiah Education Foundation, Vaddeswaram
Andhra Pradesh, India

Bidita Khandelwal
Department of General Medicine
Sikkim Manipal Institute of Medical Sciences, Sikkim Manipal University
Gangtok, Sikkim, India

Alex Khang
Department of Artificial Intelligence
Global Research Institute of Technology and Engineering
Fort Raleigh City, NC, USA

P. Pavan Kumar
Department of Computer Science and Engineering, Koneru Lakshmaiah Education Foundation, Hyderabad
Telangana, India

Prakash Kuppuswamy
Department of Computer Science and Engineering, SRM University
Delhi-NCR, Sonepat, India

Maad M. Mijwil
Computer Techniques Engineering Department, Baghdad College of Economic Sciences University
Baghdad, Iraq

M. Mohan
Department of Computer Science and Engineering, SRM University
Delhi-NCR, Sonepat, India

Sunpreet Kaur Nanda
Electronics and Telecommunication Engineering Department
P.R. Pote Patil College of Engineering and Management
Sant Gadge Baba Amravati University
Amravati, India

Rakesh Nayak
Department of CSE
OP Jindal University
Raigarh, India

O. Obulesu
Department of Computer Science and Engineering, Koneru Lakshmaiah Education Foundation, Hyderabad
Telangana, India

Ranjit Panigrahi
Department of Computer Applications, Sikkim Manipal Institute of Technology
Sikkim Manipal University, India

Ameya Patil
Faculty of School of Business
MIT World Peace University
Pune, India

Balla Adi Narayana Raju
School of Electronics and Electrical Engineering, Lovely Professional University
Phagwara, India

Ch. Rajaramesh
Department of CSE, Vignan Institute of Technology and Science
Telangana, India

Sita Rani
Department of Computer Science and Engineering, Guru Nanak Dev Engineering College, Ludhiana
Punjab, India

Shantanu Saha
Faculty of Ramcharan School of Leadership, MIT World Peace University, Pune, India

Santosh Kumar Satapathy
Information and Communication Technology, Pandit Deendayal Energy University, Gandhinagar
Gujarat, India

Rajeev Sengupta
Faculty of School of Business
Dr. Vishwanath Karad MIT World Peace University, Pune, India

Namra Shah
Information and Communication Technology, Pandit Deendayal Energy University, Gandhinagar
Gujarat, India

Potala Venkata Subbaiah
Department of Marketing
GITAM School of Business, GITAM
(Deemed to be University)
Visakhapatnam
Andhra Pradesh, India

Suman Lata Tripathi
School of Electronics and Electrical
Engineering, Lovely Professional
University, Phagwara, India

Vivek Upadhyaya
Department of Electrical & Electronics
Engineering, Poornima University
Jaipur, India

N. Venkateswarulu
Department of Computer Science and
Engineering, Koneru Lakshmaiah
Education Foundation, Hyderabad
Telangana, India; Department of
Computer Science and Engineering
G. Narayanamma Institute of
Technology and Science, Hyderabad
Telangana, India

Lucia Vilcekova
Faculty of Management, Comenius
University Bratislava, Bratislava
Slovakia

Preface

In the dynamic landscape of contemporary data analytics, the intelligent techniques have been imperatively integrated for unlocking the full potential of predictive insights. This edited volume, titled *Intelligent Techniques for Predictive Data Analytics*, emerges as a comprehensive compendium that navigates the intricacies of data mining and predictive data analysis across diverse domains, including healthcare, agriculture, and other application areas.

The exponential growth of data in today's digital age has spurred an unprecedented demand for advanced analytical methodologies. Within this context, the chapters in this book collectively serve as a guide through the vast terrain of intelligent techniques, offering both foundational principles and innovative applications. The collaborative efforts of experts from various fields have culminated in a rich tapestry of knowledge aimed at advancing the state of the art in predictive analytics. The book extends its reach into various application areas, showcasing the versatility of intelligent techniques; from healthcare to agriculture, from financial forecasting to customer relationship management, from fraud detection to energy consumption prediction, the chapters provide a panoramic view of the impact of predictive analytics across industries.

The book is structured into 12 chapters, each dedicated to a specific facet of intelligent techniques in predictive data analytics.

The journey begins with Chapter 1 focused on various data mining techniques implemented for the intelligent predictive techniques in various application areas as covered by different chapters of this work and beyond, following which an exploration of the challenges in building predictive models and potential solutions to address them is presented in Chapter 2.

Chapter 3 explores the realm of AI-driven Digital Twins and their indispensable role within the Industry 4.0 ecosystem.

Chapter 4 presents the use of predictive data techniques in the field of healthcare, elucidating how intelligent techniques can revolutionize diagnosis, treatment planning, and patient outcomes. Following which, Chapter 5 focuses on accurate interpretation of physiological signals associated with sleep disorders based on the signal waveform for automated sleep stage scoring systems to diagnose sleep disorders.

Chapter 6 presents the use of predictive analytics for marketing and sales of products using smart trolley with automated billing system in shopping malls.

Chapter 7 explores the predictive analytics, paired with relevant financial data, technical indicators, and advanced statistical approaches, as a potent tool for improving stock market forecasting.

Chapter 8 discusses the escalating challenges in cybersecurity due to the surge in data, communication systems, and cyber threats. The goal is to enhance cybersecurity by leveraging intelligent AI systems for dynamic decision-making and automated security updates in the face of evolving cyber threats.

Chapters 9 and 10 together explore the integration of cutting-edge technologies with conventional farming methods for precision agriculture with fine-grid crop management. Large amount of data produced with smart sensors is used intelligently for plant disease prediction and providing controlled, data-driven insights to farmers to manage decisions that optimize resource utilization, increase yields, and promote sustainable farming methods.

Chapter 11 sheds light on the use of intelligent techniques for predictive maintenance and load forecasting to optimize operations, enhance reliability, and save maintenance costs by proactively scheduling maintenance tasks in various industries.

Forecasting the likelihood of a future event occurring in several facets of a company's operations, including customer churn, loan default, staff performance, and so on, can save another big cost for an industry. Lastly, Chapter 12 targets to study the use of predictive analytics tools for predicting employee turnover and to classify the employee group into stayer and leaver so that effective planning of human resource can be made.

As editors, our goal has been to curate a collection that not only reflects the current landscape of intelligent techniques but also anticipates future trends and challenges.

We believe that this book will be an invaluable resource for researchers, professionals, and students seeking to deepen their understanding of predictive data analytics and harness the power of intelligent techniques. The interdisciplinary nature of the content ensures that readers from diverse backgrounds will find relevant insights, fostering a holistic appreciation for the transformative potential of predictive analytics in our data-driven world.

Happy reading!
Editors

Acknowledgments

As editors, we extend our thanks with profound gratitude to the multitude of individuals who have contributed to the creation of this edited volume.

First and foremost, we express our deepest appreciation to the esteemed authors whose expertise, dedication, and willingness to share their knowledge have been pivotal in shaping this collection into a comprehensive resource.

Editors are indebted to their respective organizations, colleagues, and mentors whose unwavering support and invaluable guidance have been instrumental throughout every stage of this project. Your insightful feedback, encouragement, and scholarly counsel have undoubtedly elevated the quality of this work.

We extend our sincere gratitude to the team at Wiley for their professionalism, expertise, and tireless efforts in bringing this project to fruition. From initial concept to final publication, your meticulous attention to detail and commitment to excellence have been evident at every turn.

Special recognition is also due to the reviewers whose constructive feedback and thoughtful critiques have helped refine and strengthen the content of this volume. Your expertise and dedication to advancing scholarly discourse are deeply appreciated.

To our families and loved ones, we owe a debt of gratitude for your unwavering patience, understanding, and encouragement throughout the countless hours spent immersed in this project. Your steadfast support has been a source of strength and inspiration.

Lastly, we dedicate this book to the readers—the students, scholars, practitioners, and enthusiasts—who will engage with its contents. May the book serve as a beacon of knowledge, sparking curiosity, fostering critical thinking, and inspiring further inquiry in the years to come.

Editors

1

Data Mining for Predictive Analytics

Prakash Kuppuswamy[1], *Mohd Dilshad Ansari*[1], *M. Mohan*[1], *and Sayed Q.Y. Al Khalidi*[2]

[1] *Department of Computer Science and Engineering, SRM University Delhi-NCR, Sonepat, India*
[2] *Department of Information Technology, King Khalid University, Abha, Kingdom of Saudi Arabia*

1.1 Introduction

Data mining is the process of discovering meaningful patterns, correlations, and relationships in large datasets using computational algorithms [1–3]. It involves various steps, including data preprocessing, pattern discovery, and result interpretation, extracting knowledge, and making informed decisions from large and complex datasets [4–6]. It combines techniques from various domains such as statistics, machine learning, artificial intelligence, and database systems to explore and analyze structured and unstructured data to gain insights and generate actionable information [7–9]. This chapter aims to delve into the concept of data mining, its methods, applications, challenges, and future research scope.

The primary goals of data mining are prediction, classification, association, clustering, and anomaly detection [10, 11]. These techniques help businesses and researchers derive valuable insights, support decision-making processes, and uncover hidden patterns within data [12, 13]. In Figure 1.1, the process of data mining typically includes data collection, data preprocessing, feature selection, algorithm selection, pattern discovery, interpretation, and evaluation. Each step plays a crucial role in extracting meaningful information from the data [14–16].

Intelligent Techniques for Predictive Data Analytics, First Edition.
Edited by Neha Singh, Shilpi Birla, Mohd Dilshad Ansari, and Neeraj Kumar Shukla.

Published 2024 by John Wiley & Sons, Inc.

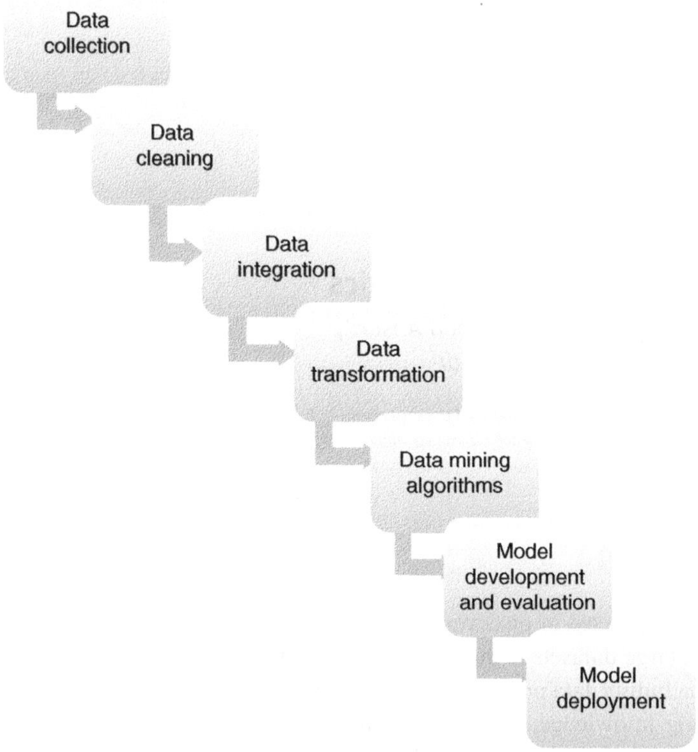

Figure 1.1 Stages of data mining process.

1.1.1 Data Mining Process

Data mining involves a systematic process to extract knowledge or patterns from large datasets. The data mining process typically consists of the following stages:

a) **Data Collection:** Collecting relevant and comprehensive datasets is crucial for accurate predictions. This data can be obtained from internal databases, third-party sources, or through data scraping techniques [17, 18].
b) **Data Cleaning:** Raw data often contains inconsistencies, missing values, and anomalies. Data cleaning involves removing or correcting these errors to ensure the accuracy and reliability of the data [19, 20].
c) **Data Integration:** In many cases, data is obtained from multiple sources and requires to be integrated into a unified dataset. This stage involves consolidating and merging data from different sources to establish a comprehensive dataset for analysis [21, 22].
d) **Data Transformation:** Data transformation involves converting raw data into a suitable format for analysis. This process includes data normalization, standardization, and feature engineering [13, 23].

e) **Data Mining Algorithms:** Selecting the appropriate algorithms based on the nature of the data and the desired outcome is crucial. Various data mining algorithms such as decision trees, neural networks, and association rules can be employed for predictive analytics [24, 25].
f) **Model Development and Evaluation:** Once the data mining algorithms have been applied, models are developed using historical data. These models are subsequently evaluated using performance metrics to assess their accuracy and effectiveness [26, 27].
g) **Model Deployment:** Once the models are considered reliable, they can be deployed in real-world scenarios to make predictions based on new data [28, 29].

Data analytics tools encompass a range of software and platforms designed to enable the extraction, analysis, and visualization of data. These tools utilize various techniques, including data mining, statistical analysis, machine learning, and visualization, to transform raw data into actionable insights. Data analytics tools come with a variety of features that enable efficient data mining, including data integration, data preprocessing, exploratory data analysis, statistical modeling, machine learning algorithms, and visualization capabilities. Using data analytics tools for data mining offers several advantages, including improved decision-making, enhanced productivity, increased efficiency, greater accuracy, and the ability to uncover hidden patterns and trends in data [12, 30, 31]. The most popular data mining techniques are as follows:

- Classification
- Clustering
- Association Rule Mining
- Regression
- Anomaly Detection

1.2 Background Study

There is a great deal of data being generated worldwide in the modern era, which is the age of data. An analysis of small, limited-dimensional data can be conducted using a basic computer. In the medical sciences, the utilization of data mining techniques has emerged as an emerging trend. However, traditional text-mining approaches are not adequate to manage the current surge in published data [32]. To find hidden features and correlations in large volumes of data, artificial intelligence-based text mining tools are being developed and used. In the field of industrial data mining, large scale data and predictive analytics present the greatest challenges [33]. In the context of extracting and analyzing underlying knowledge from a data repository, it is necessary to use data mining and machine learning techniques due to the rapid changes in the industrial environment and the availability of huge customer data. Better administration will enable better decisions in the future to manage resources and improve efficiency [34]. With the aim of creating business value, business analytics enable organizations to make better, quicker, and more intelligent decisions. At present, descriptive and predictive analytics are dominant in academic and industrial research. The field of prescriptive analytics, which seeks to determine the best course of action, has been attracting more and more attention in recent years. Often, predictive analytics is regarded as the next step in improving business performance through improved decision-making ahead of time [35]. *Data mining technique that gets insights into what has happened in the past using descriptive analytics. Forecasting and statistical techniques are used in predictive analytics to know what's likely to happen in the future. A predictive analytics technique uses algorithms to determine what actions should be taken to change what will likely happen in the future.* Starting with an introduction to analytics, it examines data mining and the role of machine learning and artificial intelligence in analytics. Comparisons are drawn based on the purposes for which the techniques are used. In addition to having advantages and disadvantages, these techniques are unique in their own way.

1.3 Applications of Data Mining

The data mining process has transformed decision-making processes across various industries, enabling organizations to anticipate trends, make accurate predictions, and optimize their operations – from marketing and healthcare to finance and supply chain management. With the ongoing advancements in technology, data mining is poised to have an even more profound impact on decision-making processes, unlocking new opportunities, and driving innovation across sectors in the foreseeable future.

1.3.1 Healthcare Industry

Data mining plays a crucial role in the healthcare sector, where the analysis of patient records, medical images, clinical trial data, and genomic data provides valuable insights. Also, it plays a vital role in disease diagnosis, patient monitoring, drug discovery, and personalized medicine. It assists healthcare professionals identify correlations between symptoms, predict disease outcomes, and optimize treatment plans. The key applications of data mining in healthcare include the following:

1) **Disease Prediction and Diagnosis:** By analyzing large-scale patient records, data mining algorithms can identify patterns, risk factors, and early indicators of diseases. This helps in predicting disease outcomes, improving diagnostic accuracy, and facilitating personalized treatments.
2) **Drug Discovery and Development:** Data mining techniques can analyze vast databases of chemical compounds to identify potential drug candidates,

accelerating the drug discovery process by reducing the time and cost involved in traditional laboratory-based approaches.

3) **Healthcare Management and Cost Reduction:** Data mining enables healthcare providers to analyze patient data, optimize resource allocation, and identify potential areas for cost reduction. It helps in streamlining operations, improving efficiency, and enhancing patient care.

1.3.2 Finance and Banking

Data mining applications in the finance and banking sectors have revolutionized the way financial institutions operate. The utilization of vast amounts of financial data combined with data mining techniques offers several advantages:

1) **Fraud Detection:** Data mining algorithms can identify patterns of fraudulent activities, such as credit card fraud, money laundering, or identity theft. By analyzing transactional data, anomalies, and suspicious patterns can be detected, enabling timely intervention.
2) **Customer Segmentation and Personalized Marketing:** By analyzing customer data, data mining enables financial institutions to segment customers into different groups based on their behavior, preferences, and needs. This helps in creating targeted marketing campaigns and personalized offers, enhancing customer satisfaction and loyalty.
3) **Risk Analysis and Assessment:** Data mining techniques help financial institutions in assessing credit risk, predicting loan defaults, and identifying potential investment opportunities. By analyzing historical data and market trends, data mining algorithms assist in making informed decisions and reducing financial risks.

1.3.3 Retail and E-commerce

Data mining plays a key role in the retail and e-commerce sectors, where vast amounts of customer data are generated. It helps businesses gain insights into customer behavior, buying patterns, market trends, and sales forecasting. It enables businesses to make data-driven decisions, improve marketing strategies, and optimize business operations. The applications and uses of data mining in this industry include the following:

1) **Market Basket Analysis:** Data mining algorithms can analyze customer purchase patterns to identify associations and correlations among products. This helps in cross-selling, offering product recommendations, and optimizing inventory management.

2) **Customer Churn Prediction:** By analyzing customer data, data mining techniques can predict customer churn, enabling targeted retention strategies. This helps in reducing customer attrition, improving customer service, and enhancing customer satisfaction.
3) **Price Optimization and Demand Forecasting:** Data mining enables retailers to analyze historical sales data, market trends, and customer behavior to optimize pricing strategies. By predicting demand patterns, retailers can set competitive prices and maximize profitability.

1.3.4 Manufacturing and Supply Chain

Data mining is widely utilized in the manufacturing and supply chain sectors to optimize operations, improve efficiency, and reduce costs. Key applications include the following:

1) **Predictive Maintenance:** By analyzing sensor data and historical maintenance records, data mining techniques can predict equipment failures, enabling proactive maintenance. This helps in minimizing downtime, reducing maintenance costs, and optimizing production efficiency.
2) **Supply Chain Optimization:** Data mining algorithms can analyze supply chain data to identify bottlenecks, inefficiencies, and areas for improvement. This facilitates optimized inventory management, demand forecasting, and efficient distribution, leading to cost savings and enhanced customer satisfaction.
3) **Quality Control and Defect Detection:** Data mining techniques can analyze production data to identify patterns, correlations, and root causes of product defects. This enables timely interventions, improving product quality, reducing waste, and ensuring customer satisfaction.

1.3.5 Recommender Systems

Recommender systems use data mining techniques to suggest relevant products, movies, music, or news articles to users. Collaborative filtering and content-based filtering are widely employed in recommendation engines.

1.3.6 Social Media Analysis

Data mining allows the analysis of social media data to understand user preferences, conduct sentiment analysis, and execute viral marketing campaigns. It helps businesses and organizations gain insights into consumer behavior and enhance brand reputation.

1.4 Challenges of Data Analytics in Data Mining

1) **Data Quality:** Data mining heavily relies on the quality and completeness of the data. Noisy or incomplete data can lead to inaccurate results and hinder the effectiveness of data mining techniques.
2) **Data Privacy and Ethics:** Data analytics must address concerns related to data privacy, security, and ethical considerations. Organizations need to adhere to privacy regulations, implement robust security measures, and ensure responsible data handling practices.
3) **Accessibility and Interpretability:** To make data analytics accessible to a wider audience, tools and techniques should be easy to understand and use. Interpretable models are also essential to gain user trust, ensure regulatory compliance, and explain the reason behind decisions.
4) **Data Quality and Integration:** Data analytics heavily rely on the quality and integration of data from various sources. Organizations need to invest in data governance, data cleansing, and data integration to ensure accurate and reliable analytics results.
5) **Scalability and Computing Power:** As data continues to grow exponentially, data analytics techniques need to be scaled efficiently. Distributed computing platforms, cloud computing, and high-performance computing will play crucial roles in handling large-scale data analytics tasks.
6) **Explainable AI and Fairness:** To gain user trust and meet regulatory requirements, data analytics techniques need to be explainable and fair. Researchers are actively exploring techniques to make AI models more transparent, interpretable, and free from bias.

1.5 Significance of Data Analytics Tools for Data Mining

1) **Business Intelligence and Decision-making:** Data analytics tools assists businesses gain insights from their data, enabling informed decision-making. These tools facilitate market analysis, customer segmentation, sales forecasting, and identification of trends and patterns in business data.
2) **Fraud Detection and Risk Management:** Data analytics tools play a crucial role in detecting fraudulent activities and managing risks. They analyze large amounts of data to identify unusual patterns and outliers, help organizations prevent fraud, analyze risks, and ensure compliance.
3) **Healthcare and Medical Research:** In the healthcare industry, data analytics tools aid in diagnosis, patient monitoring, drug discovery, and personalized medicine. These tools analyze medical records, clinical data, and genomic

information to identify correlations, predict disease outcomes, and support medical research.

4) **Social Media Analysis:** Data analytics tools help in analyzing social media data to understand user preferences, sentiment analysis, and brand reputation management. These tools mine social media platforms to identify influencers, detect trends, and assess the effectiveness of marketing campaigns.
5) **Cyber Security:** Data analytics tools are instrumental in detecting and preventing cyber threats. They analyze network traffic, system logs, and user behavior patterns to identify anomalies, detect intrusion attempts, and enhance security measures.
6) **Telecommunications:** Data analytics tools assist telecommunication companies in analyzing customer data, call records, and network performance to optimize network capacity, improve service quality, and identify customer churn patterns.

1.6 Life Cycle of Data Analytics

Predictive analytics has become an indispensable tool for organizations seeking to gain a competitive edge. It enables businesses to extract valuable insights from their data, identify patterns, and make accurate predictions about future events and outcomes. However, the process of leveraging predictive analytics is not as straightforward as it may seem. It requires a well-defined and structured approach to ensure the success and effectiveness of the analyses. In this chapter, we will explore the life cycle of predictive analytics, which consists of various stages and tasks that organizations must navigate to unlock the power of data-driven decision-making. Figure 1.2 describes about life cycle process of the data analytics model.

1.6.1 Defining the Problem

The initial step in the predictive analytics life cycle is defining the problem or question that the organization aims to address. This involves understanding the business objectives, identifying key performance indicators (KPIs), and determining which specific insights or predictions will drive value for the organization. Defining the problem sets the foundation for the subsequent stages of the life cycle and guides the selection of appropriate data and methods.

1.6.2 Data Collection and Preparation

The success of predictive analytics heavily relies on the quality and relevance of the data used. In this stage, organizations gather the necessary data from various sources, which may include internal databases, external sources, public repositories,

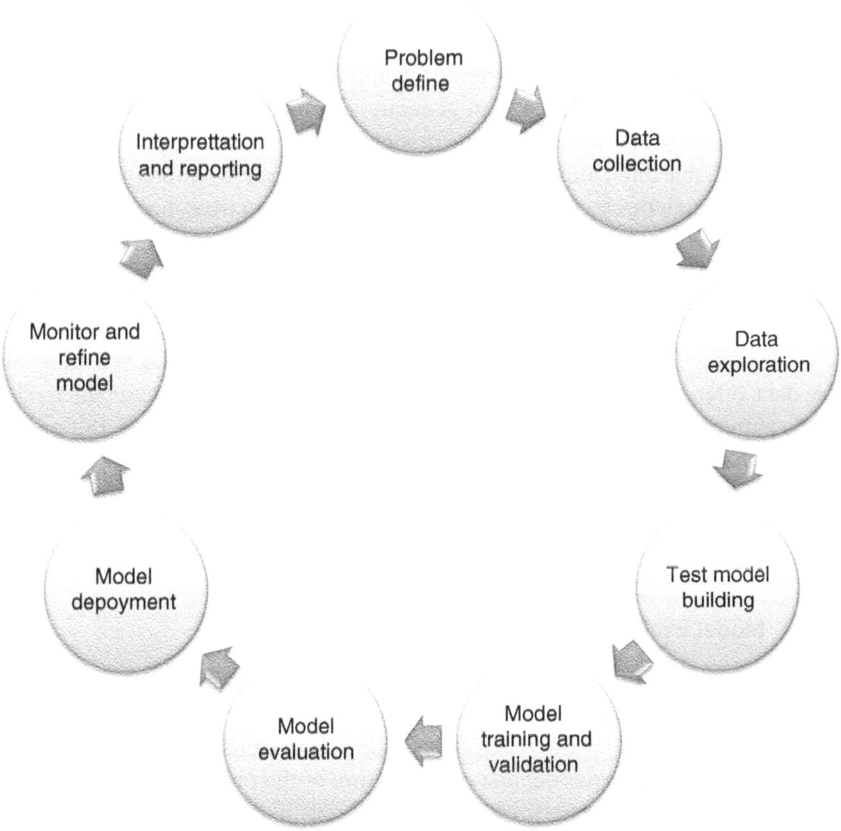

Figure 1.2 Life cycle of predictive analytics.

or customer feedback. The collected data may be structured, unstructured, or semi-structured. Once the data is gathered, it needs to be cleaned, transformed, and prepared for analysis. This involves tasks such as handling missing values, removing outliers, standardizing formats, and ensuring data integrity.

1.6.3 Data Exploration and Visualization

In this stage, organizations explore the collected data to gain a comprehensive understanding of its characteristics, patterns, and relationships. Through descriptive statistics, data profiling, and visualization techniques, analysts uncover hidden insights, identify trends, and detect anomalies. Data exploration helps in feature selection, as it allows analysts to determine which variables are relevant to the predictive model and should be included in further analyses.

1.6.4 Model Development

Once the data is prepared and understood, organizations move on to the model development stage. This involves selecting appropriate algorithms or techniques to build predictive models that can generate accurate predictions based on the desired outcomes. The selection of the models depends on the nature of the problem, the available data, and the required level of accuracy. Common models used in predictive analytics include regression models, decision trees, neural networks, and support vector machines.

1.6.5 Model Training and Validation

After choosing the model, organizations need to train it using a subset of the available data, often referred to as the training set. During the training phase, the model learns patterns, relationships, and rules from the data. Once trained, the model needs to be validated to assess its predictive performance. This involves using another subset of the data, the validation set, to evaluate the model's accuracy and ensure that it is not overfitting the training data. Strategies such as cross-validation and hold-out validation are used to assess the model's generalization ability.

1.6.6 Model Evaluation and Selection

At this stage, organizations evaluate the performance of different models based on predefined evaluation metrics. Common metrics include accuracy, precision, recall, F1 score, and area under the receiver operating characteristic (ROC) curve. The selected evaluation metrics depend on the nature of the problem such as classification, regression, or time series forecasting. The evaluation process helps organizations identify the most suitable model to deploy for their specific needs.

1.6.7 Model Deployment and Integration

Once a preferred model is selected, it is deployed and integrated into the organization's existing systems or processes. This can involve integrating the model within a software application, a website, or an automated decision-making system. The model should be implemented in a production environment where it can receive real-time data inputs and provide predictions or recommendations. It is crucial to monitor the performance of the deployed model, ensuring that it remains accurate and effective over time.

1.6.8 Model Monitoring and Maintenance

Predictive models are not static entities; they require ongoing monitoring and maintenance to ensure their continued effectiveness. Organizations need to continuously track the performance of the model, monitor inputs and outputs, evaluate its accuracy, recalibrate when necessary, and update the model as new data becomes available. Monitoring helps identify any drift or degradation in the model's performance and allows for timely adjustments to maintain its predictive power.

1.6.9 Interpretation and Reporting

The final stage of the predictive analytics life cycle involves interpreting the model's results and reporting them to relevant stakeholders. The predictions and insights generated by the model need to be communicated in a clear and understandable manner. This can involve creating visualizations, dashboards, or reports that effectively convey the predictions and their implications. Interpretation and reporting ensure that decision-makers can understand and act upon the predictions to drive data-driven decision-making.

The life cycle of predictive analytics is a structured approach that guides organizations through the stages of problem definition, data collection and preparation, model development, training, evaluation, and deployment. Each stage plays a crucial role in ensuring the accuracy, validity, and effectiveness of the predictive models. By following this life cycle, organizations can harness the power of data-driven decision-making, gain insights into future events, and improve their business strategies, operations, and overall performance.

1.7 Predictive Analytics Model

Predictive analytics has emerged as a transformative technology that enables organizations to make accurate forecasts and predictions about future events and outcomes. By analyzing historical data and applying various statistical and machine learning techniques, predictive analytics helps organizations gain insights, identify trends, and make informed decisions. In this chapter, we will explore the various predictive analytics methods and their applications across different industries.

1.7.1 Regression Analysis

Regression analysis is one of the most widely used predictive analytics methods. It involves analyzing the relationship between a dependent variable and one or more independent variables. By fitting a mathematical model to historical data, regression analysis enables organizations to predict future values of the dependent variable based on the independent variables. This method is particularly useful for forecasting sales, demand, and pricing.

1.7.2 Time Series Analysis

Time series analysis focuses on analyzing data points collected at regular intervals over time. It aims to identify patterns, trends, and seasonality in the data and use them to make accurate future predictions. Time series forecasting techniques, such as moving averages, exponential smoothing, and autoregressive integrated moving average (ARIMA) models, help organizations predict future values based on historical patterns. Time series analysis finds wide application in finance, sales forecasting, and demand planning.

1.7.3 Classification

Classification is a machine learning technique that involves predicting the category or class of a given observation based on a set of input features. It builds a model using historical data with known classifications and applies that model to predict the class labels of new, unseen data. Classification is used in various applications such as customer segmentation, fraud detection, and sentiment analysis. Popular classification algorithms include logistic regression, decision trees, support vector machines, and random forests.

1.7.4 Clustering

Clustering is a technique used in predictive analytics to group similar data points together based on their similarities or distances. It helps in identifying hidden patterns and structures within datasets without predefined class labels. Clustering algorithms, including k-means clustering and hierarchical clustering, enable organizations to discover valuable insights for customer segmentation, market research, and anomaly detection.

1.7.5 Neural Networks

Neural networks are a type of machine learning algorithm inspired by the human brain's structure and function. They consist of interconnected artificial neurons that learn from historical data to make predictions or decisions. Neural networks are particularly powerful when dealing with complex, nonlinear relationships in data. They find applications in diverse areas such as image and speech recognition, fraud detection, and predictive maintenance. Figure 1.3 outlines predictive analytics models and algorithms.

1.7.6 Decision Trees

Decision trees are a popular predictive analytics method that visually represents decisions and their possible consequences. This method involves creating a

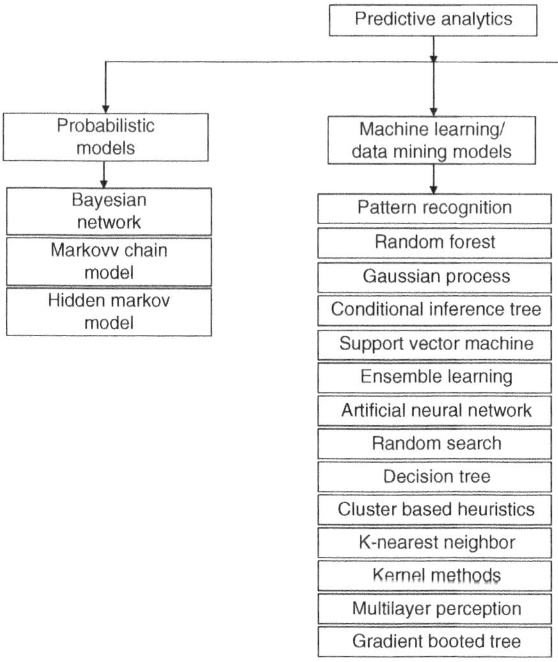

Figure 1.3 Predictive algorithms.

tree-like structure where each internal node represents a decision based on an attribute, and each leaf node represents an outcome or class label. Decision trees are easy to interpret and understand, making them suitable for risk assessment, customer churn prediction, and diagnosis in healthcare.

1.7.7 Ensemble Methods

Ensemble methods combine predictions from multiple individual models to create a more accurate and robust prediction. Bagging, boosting, and stacking are commonly used ensemble techniques. Bagging uses bootstrap aggregation to generate diverse models while boosting focuses on iteratively improving the performance of weak models. Stacking combines the predictions of multiple models using a meta-learner. Ensemble methods are widely applied in credit scoring, stock market predictions, and anomaly detection.

1.7.8 Text Analytics

Text analytics, also known as text mining or natural language processing (NLP), is concerned with extracting meaning and insights from unstructured text data.

It involves techniques such as sentiment analysis, topic modeling, and text classification to analyze large volumes of text and make predictions or identify trends. Text analytics find applications in social media analysis, customer feedback analysis, and market research.

1.7.9 Anomaly Detection

Anomaly detection involves identifying unusual or rare events, patterns, or data points that deviate significantly from the norm. It helps organizations detect fraud, network intrusions, equipment malfunctions, or any other unexpected behavior. Various statistical techniques, such as Z-score, interquartile range (IQR) and modified z-scores.

1.7.10 Collaborative Filtering

Collaborative filtering is a technique commonly used in recommendation systems to predict a user's preferences or interests based on the preferences of similar users or items. It leverages historical data on user–item interactions to make personalized recommendations. Collaborative filtering is widely employed in e-commerce, social media, and streaming platforms to provide users with personalized suggestions and enhance customer experience.

Predictive analytics methods have revolutionized decision-making processes across industries by enabling organizations to see into the future, anticipate trends, and make informed strategic choices. From regression analysis and time series forecasting to clustering, neural networks, and ensemble methods, the range of predictive analytics techniques continues to expand as data becomes increasingly available and sophisticated algorithms emerge. By harnessing the power of predictive analytics, organizations can optimize their operations, predict customer behavior, mitigate risks, and gain a competitive edge in today's data-driven environment.

1.8 Data Analytics Tools

Data is considered the new oil in today's digital era. Organizations are inundated with vast amounts of data from various sources. To extract valuable insights and gain a competitive edge, they turn to predictive analytics, a branch of data analytics that uses historical and real-time data to forecast future trends and outcomes. By leveraging advanced statistical models and algorithms, predictive analytics empowers decision-makers to make informed choices, mitigate risks, and optimize opportunities. This section explores the expansive uses of predictive analytics tools and their transformative impact on a wide range of applications.

1.8.1 RapidMiner

RapidMiner is a popular open-source data analytics tool used for data mining and predictive analytics. It provides a user-friendly interface with drag-and-drop functionality, making it accessible to users with varying technical expertise. Some key features of RapidMiner include the following:

a) **Data Preprocessing:** RapidMiner provides a wide range of data preprocessing functionalities such as data cleaning, filtering, and feature selection. It enables users to handle missing data, normalize variables, and perform other data transformation tasks.
b) **Visualization:** RapidMiner offers a variety of visualization tools to visually explore and understand patterns within the data. This includes scatter plots, histograms, decision trees, and more, enabling users to gain insights from their data intuitively.
c) **Machine Learning Algorithms:** RapidMiner supports various machine learning algorithms such as decision trees, neural networks, clustering, and association rule mining. Users can apply these algorithms to build predictive models and make informed predictions.
d) **Deployment and Integration:** RapidMiner makes it easy to deploy models into production environments. It provides APIs and integration options to seamlessly integrate with other systems and tools.

RapidMiner finds applications in industries such as healthcare, finance, marketing, and manufacturing, enabling organizations to gain actionable insights from their data.

1.8.2 Konstanz Information Miner (KNIME)

KNIME is an open-source data analytics platform that provides a graphical interface for building data workflows. It offers a wide range of data mining and machine learning functionalities. Some notable features of KNIME include the following:

a) **Drag-and-drop Interface:** KNIME's intuitive interface allows users to create data workflows by dragging and dropping nodes. Each node represents a specific operation or analysis, making it easy to design complex data workflows without writing code.
b) **Large Selection of Nodes:** KNIME provides a vast library of prebuilt nodes for tasks such as data cleaning, data transformation, feature selection, and model building. Users can combine these nodes to create customized data mining workflows.
c) **Integrated Analytics:** KNIME seamlessly integrates with popular analytics tools like R and Python. This allows users to leverage their preferred programming languages and libraries within the KNIME environment.

d) **Collaboration and Sharing:** KNIME enables collaboration among team members by allowing the sharing of workflows, analyses, and models. This facilitates knowledge sharing and improves efficiency.

KNIME finds applications in various domains including finance, life sciences, retail, and manufacturing.

1.8.3 Statistical Analysis System (SAS)

SAS is a comprehensive and widely used analytics platform that provides a range of tools for data mining and predictive analytics. SAS offers a rich set of features and functionalities, making it suitable for complex analytical tasks. Key features of SAS include the following:

a) **Data Management:** SAS provides data management capabilities, including data integration, data cleaning, and data transformation. It allows users to prepare large datasets for analysis and handle missing values and outliers effectively.
b) **Statistical Analysis:** SAS offers a wide range of statistical procedures for exploratory data analysis, hypothesis testing, and regression analysis. Users can apply these procedures to extract meaningful insights from the data and identify relationships.
c) **Machine Learning and Predictive Modeling:** SAS provides a comprehensive collection of machine learning algorithms that can be used for building predictive models. These algorithms include decision trees, neural networks, support vector machines, and more.
d) **Visual Analytics:** SAS offers interactive and visually appealing visual analytics tools for exploratory data analysis. Users can create compelling visualizations, dashboards, and reports to communicate insights effectively.

SAS is widely used in industries such as finance, healthcare, and government, where complex analytics and data mining are critical for decision-making.

1.8.4 Python with Scikit-learn

Python, a popular programming language, coupled with the Scikit-learn library, offers a powerful and flexible environment for data mining and predictive analytics. Python and Scikit-learn provide the following functionalities:

a) **Extensive Data Mining and Machine Learning Libraries:** Python's Scikit-learn library offers a wide range of algorithms for classification, regression, clustering, and dimensionality reduction. These include decision trees, support vector machines, random forests, and more.

b) **Integration with Other Libraries:** Python seamlessly integrates with other libraries, such as NumPy, pandas, and Matplotlib, providing a complete data analytics ecosystem. These libraries allow for efficient data manipulation, data visualization, and data preprocessing.
c) **Flexibility and Customization:** Python offers flexibility for users to customize their data mining and predictive analytics workflows. It allows users to tailor the analysis and models to meet specific requirements and solve unique challenges.
d) **Active Open-source Community:** Python has a vibrant and active community that continuously develops new packages and tools for data mining and predictive analytics. This ensures that Python remains at the forefront of data analytics tooling.

Python with Scikit-learn is widely used across industries including finance, healthcare, e-commerce, and cyber-security.

1.8.5 Programming R

Programming R is a widely used programming language in the field of data analytics. They offer numerous libraries and packages specifically designed for data mining such as Scikit-learn, TensorFlow, Keras, Pandas, NumPy, and SciPy. These libraries provide a wide range of data mining algorithms, allowing users to preprocess, analyze, and visualize data efficiently.

1.8.6 Waikato Environment for Knowledge Analysis (Weka)

Weka is a popular open-source data mining tool developed in Java. It provides a comprehensive collection of machine-learning algorithms and tools for data preprocessing, classification, regression, clustering, association rule mining, and visualization. Weka also features a user-friendly interface and supports integration with other programming languages.

1.8.7 Tableau

Tableau is a widely used data visualization tool that enables users to analyze and visualize data in an interactive and intuitive manner. It allows users to create visually appealing dashboards, charts, and graphs, providing a powerful platform for exploring data mining results and presenting findings to stakeholders effectively.

1.8.8 Microsoft Power BI

Microsoft Power BI is a business intelligence tool that enables to connect to multiple data sources, transform and clean data, and create interactive visualizations and reports. It provides powerful data analytics capabilities, including data mining algorithms, time-series analysis, forecasting, and NLP.

1.9 Benefits of Predictive Analytics Techniques

Predictive analytics has revolutionized the way organizations approach customer behavior analysis, allowing them to predictively segment customers and personalize their offerings. This section examines how predictive analytics is applied to various segmentation, churn prediction, demand forecasting, and recommendation systems, helping organizations optimize marketing strategies, etc.

1) **Enhanced Decision-Making:** Data analytics techniques enable organizations to make data-driven decisions based on evidence and insights, rather than intuition. Advanced modeling and analysis techniques provide accurate predictions, mitigate risks, and identify opportunities.
2) **Improved Operational Efficiency:** Data analytics techniques optimize resource allocation, streamline processes, and enhance efficiency. Predictive analytics helps organizations avoid stockouts or overstocking, while prescriptive analytics suggests optimal production schedules and resource allocation.
3) **Personalized Customer Experiences:** Data analytics techniques, such as recommender systems and segmentation analysis, allow organizations to personalize customer experiences. By understanding customer preferences and behavior patterns, organizations can provide targeted recommendations, customized marketing campaigns, and personalized product offerings.
4) **Fraud Detection and Risk Mitigation:** Data analytics techniques, including anomaly detection and network analysis, help identify fraudulent activities and mitigate risks. Organizations can proactively detect anomalies, investigate suspicious activities, and prevent financial losses and security breaches.
5) **Innovation and Competitive Advantage:** Data analytics techniques open avenues for innovation by uncovering emerging market trends, customer preferences, and hidden patterns. Organizations can gain a competitive edge by leveraging these insights to develop innovative products, services, and business strategies.

1.10 Applications of Predictive Analytics Model

The organizations are generating vast amounts of data at an unprecedented rate. The challenge lies in harnessing this data and transforming it into valuable insights. This is where data analytics comes into play. Data analytics is the practice of examining, cleaning, transforming, and modeling data to discover useful information, draw conclusions, and support decision-making. In this chapter, we will explore the numerous applications of data analytics across various sectors and industries.

1.10.1 Business Intelligence

Data analytics plays a crucial role in business intelligence by providing organizations with visibility into their performance, customer behavior, and market trends. By analyzing customer preferences, organizations can gain insights into consumer behavior patterns, enabling targeted marketing and product development strategies. Data analytics also helps businesses identify sales trends, optimize pricing strategies, and forecast demand, facilitating better decision-making and increasing profitability.

1.10.2 Healthcare

Data analytics has revolutionized the healthcare industry, enabling better patient care, medical research, and the identification of disease patterns. Analyzing patient data allows healthcare providers to identify trends and risk factors, enabling early intervention and disease prevention. By leveraging data analytics, medical researchers can extract insights from large datasets, leading to advancements in treatment options, drug discovery, and improved clinical outcomes.

1.10.3 Finance

Data analytics has transformed the finance industry, enabling organizations to make data-driven decisions and manage risk more effectively. By analyzing historical financial data, organizations can identify trends, patterns, and irregularities, enabling fraud detection, risk management, and investment decision-making. Real-time data analytics also allow financial institutions to monitor transactions, detect suspicious activities, and promptly respond to market changes.

1.10.4 Supply Chain Management

Data analytics play a vital role in optimizing supply chain management by improving inventory management, logistics, and demand forecasting. By analyzing historical purchasing data, organizations can predict future demand, optimize inventory levels, and enhance procurement processes. Data analytics also enables organizations to track and monitor supplier performance, identify bottlenecks, and improve overall operational efficiency.

1.10.5 Manufacturing

Data analytics is transforming the manufacturing industry by facilitating predictive maintenance, quality control, and process optimization. By analyzing real-time sensor data from machinery, organizations can predict and prevent equipment failures, reducing downtime and maintenance costs. Data analytics also enables organizations to identify quality issues in real time, reducing product defects and improving customer satisfaction. In addition, data analytics helps manufacturers optimize production processes, minimize waste, and maximize overall productivity.

1.10.6 Marketing and Advertising

Data analytics has changed the face of marketing and advertising, allowing organizations to target their audiences more effectively and measure the success of their campaigns. By analyzing customer data, organizations can identify their target audience's demographics, preferences, and purchasing behavior, enabling personalized marketing campaigns. Data analytics also provides marketers with valuable insights into the effectiveness of various marketing channels and helps allocate advertising budgets more efficiently.

1.10.7 Smart Cities

Data analytics plays a vital role in the development of smart cities, enabling improved urban planning, resource management, and citizen services. By analyzing data from sensors, satellites, and social media, government authorities can collect valuable insights into traffic patterns, air quality, energy consumption, and public services. This data-driven approach allows authorities to optimize resource allocation, reduce congestion, and enhance the quality of life for citizens.

The applications of data analytics are vast and evolving rapidly, transforming industries and enabling organizations to make data-driven decisions. From business intelligence and healthcare to finance, supply chain management, manufacturing, marketing, and smart cities, data analytics has become an indispensable tool for organizations seeking to gain a competitive edge. As data continues to grow exponentially, the importance of data analytics in extracting valuable insights will only increase.

1.11 Conclusion

Data mining has become an indispensable tool across various industries and sectors due to its ability to extract valuable insights from large volumes of data. Data analytics model have become essential for data mining tasks, enabling organizations to extract valuable insights from large and complex datasets. These tools offer a wide range of functionalities including data preprocessing, analysis, visualization, and modeling. Data mining for predictive analytics is a powerful tool that enables businesses and organizations to extract valuable insights from vast amounts of data. By implementing a systematic data mining process, organizations can leverage historical data to build predictive models that aid in strategic decision-making across various industries. Leveraging data mining techniques offers numerous benefits, such as improved decision-making, competitive advantage, and the identification of hidden patterns. However, there are challenges to overcome, including data accessibility, scalability, and privacy concerns. As businesses continue to generate and collect massive amounts of data, the integration of data mining with predictive analytics will become increasingly crucial in gaining a competitive edge and driving innovation. Data analytics tools for data mining play a vital role in extracting valuable insights from large datasets. These tools enable organizations across industries to leverage data-driven insights, optimize decision-making processes, and gain a competitive advantage in today's data-centric world. The future of data analytics tools holds promise for improving decision-making, enhancing efficiency, and enabling organizations to leverage the power of data to gain a competitive edge.

References

1 Gupta, M.K. and Chandra, P. (2020). A comprehensive survey of data mining. *International Journal of Information Technology* 12 (4): 1243–1257.

2 Yan, H., Yang, N., Peng, Y., and Ren, Y. (2020). Data mining in the construction industry: present status, opportunities, and future trends. *Automation in Construction* 119: 103331.

3 Arumugam, K., Naved, M., Shinde, P.P. et al. (2023). Multiple disease prediction using Machine learning algorithms. *Materials Today: Proceedings* 80: 3682–3685.

4 Johnson, M., Albizri, A., Harfouche, A., and Fosso-Wamba, S. (2022). Integrating human knowledge into artificial intelligence for complex and ill-structured problems: informed artificial intelligence. *International Journal of Information Management* 64: 102479.

5 Sharmin, S., Ahammad, T., Talukder, M.A., and Ghose, P. (2023). A hybrid dependable deep feature extraction and ensemble-based machine learning approach for breast cancer detection. *IEEE Access* 11: 87694–87708.

6 Fan, C., Chen, M., Wang, X. et al. (2021). A review on data preprocessing techniques toward efficient and reliable knowledge discovery from building operational data. *Frontiers in Energy Research* 9: https://doi.org/10.3389/fenrg.2021.652801.

7 Kumar, P., Tveritnev, A., Jan, S.A., and Iqbal, R. (2023). Challenges to opportunity: getting value out of unstructured data management. *SPE Gas & Oil Technology Showcase and Conference*, Dubai, UAE (14 March 2023).

8 Niu, Y., Ying, L., Yang, J. et al. (2021). Organizational business intelligence and decision making using big data analytics. *Information Processing & Management* 58 (6): 102725.

9 Adnan, K., Akbar, R., Khor, S.W., and Ali, A.B.A. (2020). Role and challenges of unstructured big data in healthcare. In: *Data Management, Analytics and Innovation. Advances in Intelligent Systems and Computing*, vol. 1042 (ed. N. Sharma, A. Chakrabarti, and V. Balas), 301–323. Singapore: Springer. https://doi.org/10.1007/978-981-32-9949-8_22.

10 Liu, X., Ding, Y., Tang, H., and Xiao, F. (2021). A data mining-based framework for the identification of daily electricity usage patterns and anomaly detection in building electricity consumption data. *Energy and Buildings* 231: 110601.

11 Phillips-Wren, G., Daly, M., and Burstein, F. (2021). Reconciling business intelligence, analytics and decision support systems: more data, deeper insight. *Decision Support Systems* 146: 113560.

12 Dogan, A. and Birant, D. (2021). Machine learning and data mining in manufacturing. *Expert Systems with Applications* 166: 114060.

13 Bharadiya, J.P. (2023). The role of machine learning in transforming business intelligence. *International Journal of Computing and Artificial Intelligence* 4 (1): 16–24.

14 Thakkar, A. and Lohiya, R. (2023). Fusion of statistical importance for feature selection in deep neural network-based intrusion detection system. *Information Fusion* 90: 353–363.

15 Siddiqui, M.K., Morales-Menendez, R., Huang, X., and Hussain, N. (2020). A review of epileptic seizure detection using machine learning classifiers. *Brain Informatics* 7 (1): 1–8.

16 Ahmed, N., Rafiq, J.I., and Islam, M.R. (2020). Enhanced human activity recognition based on smartphone sensor data using hybrid feature selection model. *Sensors* 20 (1): 317.

17 Tunyasuvunakool, K., Adler, J., Wu, Z. et al. (2021). Highly accurate protein structure prediction for the human proteome. *Nature* 596 (7873): 590–596.

18 Shillan, D., Sterne, J.A.C., Champneys, A., and Gibbison, B. (2019). Use of machine learning to analyse routinely collected intensive care unit data: a systematic review. *Critical Care* 23 (1): 1–11.

19 Shi, X., Prins, C., Van Pottelbergh, G. et al. (2021). An automated data cleaning method for electronic health records by incorporating clinical knowledge. *BMC Medical Informatics and Decision Making* 21 (1): 1–10.

20 Faiz, T. (2019). Multi-approaches on scrubbing data for medium-sized enterprises. *2019 International Conference on Digitization (ICD)*, Sharjah, United Arab Emirates (18–19 November 2019), pp. 75–86.

21 Bayoudh, K., Knani, R., Hamdaoui, F., and Mtibaa, A. (2021). A survey on deep multimodal learning for computer vision: advances, trends, applications, and datasets. *The Visual Computer* 38 (8): 2939–2970.

22 Mavrogiorgou, A., Kiourtis, A., Perakis, K. et al. (2019). Analyzing data and data sources towards a unified approach for ensuring end-to-end data and data sources quality in healthcare 4.0. *Computer Methods and Programs in Biomedicine* 181: 104967.

23 Clarke, D.J., Jeon, M., Stein, D.J. et al. (2021). Appyters: turning Jupyter notebooks into data-driven web apps. *Patterns* 2 (3): 100213.

24 Alsalman, Y.S., Halemah, N.K. A., AlNagi, E.S., and Salameh, W. (2019). Using decision tree and artificial neural network to predict students academic performance. *2019 10th International Conference on Information and Communication Systems (ICICS)*, Irbid, Jordan (11–13 June 2019), pp. 104–109.

25 Arcinas, M.M., Sajja, G.S., Asif, S. et al. (2021). Role of data mining in education for improving students performance for social change. *Turkish Journal of Physiotherapy and Rehabilitation* 32 (3): 6519–6526.

26 Ramesh, T.R., Lilhore, U.K., Poongodi, M. et al. (2022). Predictive analysis of heart diseases with machine learning approaches. *Malaysian Journal of Computer Science* 132–148.

27 Yang, Y., Zheng, K., Wu, C., and Yang, Y. (2019). Improving the classification effectiveness of intrusion detection by using improved conditional variational AutoEncoder and deep neural network. *Sensors* 19 (11): 2528.

28 Markov, T. (2023). A holistic approach to undesired content detection in the real world. *Proceedings of the AAAI Conference on Artificial Intelligence* 37 (12): 15009–15018.

29 Bourhnane, S., Abid, M.R., Lghoul, R. et al. (2020). Machine learning for energy consumption prediction and scheduling in smart buildings. *SN Applied Sciences* 2 (2): 297.

30 Li, C., Chen, Y., and Shang, Y. (2022). A review of industrial big data for decision making in intelligent manufacturing. *Engineering Science and Technology, an International Journal* 29: 101021.

31 Sunhare, P., Chowdhary, R.R., and Chattopadhyay, M.K. (2022). Internet of things and data mining: an application oriented survey. *Journal of King Saud University - Computer and Information Sciences* 34 (6): 3569–3590.

32 Antons, D., Grünwald, E., Cichy, P., and Salge, T.O. (2020). The application of text mining methods in innovation research: current state, evolution patterns, and development priorities. *R&D Management* 50 (3): 329–351.

33 Dai, H.-N., Wong, R.C.-W., Wang, H. et al. (2019). Big data analytics for large-scale wireless networks. *ACM Computing Surveys* 52 (5): 1–36.

34 Diez-Olivan, A., Del Ser, J., Galar, D., and Sierra, B. (2019). Data fusion and machine learning for industrial prognosis: trends and perspectives towards Industry 4.0. *Information Fusion* 50: 92–111.

35 Lepenioti, K., Bousdekis, A., Apostolou, D., and Mentzas, G. (2020). Prescriptive analytics: literature review and research challenges. *International Journal of Information Management* 50: 57–70.

2

Challenges in Building Predictive Models

Rakesh Nayak[1], Ch. Rajaramesh[2], and Umashankar Ghugar[1]

[1] *Department of CSE, OP Jindal University, Raigarh, India*
[2] *Department of CSE, Vignan Institute of Technology and Science, Telangana, India*

2.1 Introduction

Predictive modeling is the process of using data to build a model that can predict outcomes that can be used in the future. It is an important aspect of data analysis, machine learning, and artificial intelligence (AI). It is a valuable tool for businesses and organizations as it can help them to make better decisions and improve their performance. It involves statistical models and algorithms to analyze data, identify patterns, and make predictions about future outcomes. Predictive modeling has applications in various industries including healthcare, finance, marketing, transportation, and many more. However, building predictive models can be a challenging task. Several factors can affect the accuracy and performance of a predictive model. These factors include the quality of the data, the choice of model, and the way in which the model is trained and evaluated. This chapter discusses some of the challenges in building predictive models. It will also provide tips on how to overcome these challenges and build models that are accurate and reliable [1]. Data acquisition is the process of gathering and collecting data from a variety of sources. It is an essential part of the predictive modeling process, as the quality of the data will directly impact the accuracy of the model. It can be challenging to acquire the data you need to build a predictive model. The data may be scattered across different sources, it may be incomplete or inaccurate, or it may be in a format that is not easy to use [2].

Intelligent Techniques for Predictive Data Analytics, First Edition.
Edited by Neha Singh, Shilpi Birla, Mohd Dilshad Ansari, and Neeraj Kumar Shukla.

Published 2024 by John Wiley & Sons, Inc.

The success of a predictive model heavily relies on the availability, quality, quantity, format, access, compatibility, consistency, and diversity of data used to train the algorithm. Data acquisition is a crucial step in the predictive modeling process because it enables us to collect data from multiple sources, clean and transform it to facilitate reliable and accurate predictions. However, data acquisition is not without its challenges given the numerous dashboards, databases, and APIs that are available to access data. This chapter aims to highlight some of the significant challenges in data acquisition for building predictive models.

2.1.1 Different Predictive Models

Yang et al. [2] has summarize 148 predictive modeling strategies used in various software engineering activities to have a better knowledge of predictive models. Table 8 in [2] indicates the number of occurrences in which various categories of prediction models and relevant studies emerge during 2009-2020. It can be seen that 27 studies used rule-based predictive modeling techniques, with JRip and RIPPER being the most commonly used rule-based models. Statistical models are grouped into four types based on their functions, with the dimensionality reduction technique (e.g., linear discriminant analysis [LDA]) being the most commonly employed, followed by the state model. The natural language processing (NLP) family includes models that are specialized for processing human languages. In our primary research, only three articles used related methodologies as predictive models. Among all learners, spanning 7 different families, the machine learner is the most commonly used predictive model, with over 100 studies using logistic regression, Naive Bayes, and support vector machine (SVM) models to address software engineering problems. The regression family includes 25 different regression algorithms, followed by the SVM and decision tree families of machine learning (ML)-based predictive modeling approaches. The SVM and decision tree families each have 12 distinct algorithms.

In addition to the aforementioned standard ML techniques, studies commonly employ instance-based algorithms like the k-nearest neighborhood (kNN) and its variants. In addition, machine learning relies heavily on the clustering method and the learning to rank (LTR) family. Table 8 in [2] (included in the Appendix) shows the widespread use of machine learners in the years following the advent of DL, suggesting that these learners are nonetheless effective in addressing appropriate problems despite their relatively late introduction.

Many research have used ensemble algorithms to address software engineering (SE) problems; broadly speaking, these may be broken down into three types: random Forest, bagging, and boosting. Over 65% of the research favored Random Forest as the predictive model over bagging and boosting. Fifty-four papers used Boosting approaches in their trials, with AdaBoost and XGBoost being the most widely used predictive modeling strategies. Twenty-three research projects on SE use the ensemble method of bagging. Most of the research that used deep learning as a predictive model was conducted between 2018 and 2020, which coincides with the rise in popularity of DL technology. This indicates that in recent years, there has been a rising tide of interest in applying deep learning to software engineering. Table 8 in [2] shows that 113 studies have used deep learning techniques (the deep neural network [DNN] family) to solve software engineering difficulties. With 24 research employing it, CNN is the most widely utilized deep learning model. Artificial neural network (ANN) and long short-term memory networks (LSTM) follow closely after. Of the papers considered, only five made use of deep belief networks (DBNs). Perceptron models were utilized in 32 of the research, mainly in the form of multi-layer perceptron.

2.1.2 Challenges Associated with Data Acquisition

There are several challenges associated with data acquisition for predictive modeling, including [3]:

- **Data Availability:** Data acquisition begins with data discovery, identifying the sources of data that might be relevant to the predictions under consideration. One of the significant challenges in data acquisition is the availability of data from specific sources. Some databases may be restricted and require administrative approvals, while others may not be accessible due to legal or privacy constraints. Availability of data also affects data diversity, which is essential in building robust predictive models. A lack of data diversity may cause the model to make inaccurate predictions.
- **Data Quality:** One of the most significant challenges in building predictive models is data quality. In other words, data quality, or the level of trustworthiness in the data, remains a substantial challenge in data acquisition. The quality of the data affects the accuracy and reliability of the predictive model. Datasets obtained from various sources may include missing data, incomplete data, duplicate data, and data inconsistencies. Poor data quality can result in noisy or biased models that lack the ability to make accurate predictions. Missing data may skew the model predictions, while duplicate data may cause over-fitting. To overcome data quality challenges, we should ensure they have clean and comprehensive data sets. We can use data cleaning techniques, such as imputation, duplication removal, and outlier removal to improve data quality. In addition, we can use data quality metrics, such as completeness, accuracy, consistency, and timeliness, to evaluate data quality. Collecting complete and consistent data in sufficient quantities increases the likelihood of building a reliable predictive model.
- **Data Volume:** The amount of data required for training a predictive model may require substantial effort in data acquisition. The amount of data required

to build accurate models can be very large. Large datasets also require significant storage, processing power, and network bandwidth. Therefore, it is essential to invest in data acquisition solutions that can handle big data requirements.

- **Data Format:** The data must be in a format that can be used by the predictive modeling algorithm. This may require transforming the data into a different format, which can be a time-consuming and challenging task. One of the main challenges in data acquisition is dealing with diverse data formats. When collecting data from different sources, it's common for each dataset to have its unique structure, naming conventions, date formats, and more. This diversity in data formats can make cleaning and integrating data extremely challenging and time-consuming. In addition, when we train a predictive model, it is critical to ensure that the input data is in the correct format and that any necessary transformations are performed correctly. Failing to address data format issues can lead to inaccurate predictions, poorly performing models, or unreliable insights. Therefore, it is essential to have a robust data acquisition strategy that incorporates cleaning and preprocessing steps to address challenges related to data format during data acquisition.
- **Data Access:** The data may be located in a variety of sources, such as databases, files, and sensors. This can make it difficult to access and collect the data. One of the main challenges in data acquisition is data access, particularly when dealing with sensitive or proprietary data. Access to some data sources may be restricted or require permissions, which can slow down the process of collecting and integrating data into a predictive model. Moreover, different data sources may require different access methods or APIs, which may pose additional challenges for teams trying to collect data from multiple sources. In addition, some data sources may have limitations on the amount of data that can be accessed or impose restrictions on the types of analyses that can be performed on the data. Therefore, establishing suitable data access protocols and ensuring that all necessary permissions and allowances are in place is crucial in addressing the challenge of data access during data acquisition.
- **Data Compatibility:** Data acquired from multiple sources may not be compatible with the preferred format of the predictive modeling tool. The processing requirements, file format, and data structure of the data source need to be harmonized to allow integration into the predictive modeling solution. This can be a time-consuming process that requires significant effort from the data acquisition professional.
- **Data Consistency:** Data acquired from various sources may have different levels of consistency, making it challenging to merge them into a single dataset. Inconsistencies often result in model errors that can lead to poor predictions. Data profiling is necessary to identify these inconsistencies, including formatting issues, invalid characters, and duplicate records.

- **Data Diversity:** One of the main challenges of diversity in data during data acquisition is finding diverse data sources. Depending on the industry or field of application, finding diverse data that accurately represents the outcome variable (the variable being predicted) can be quite difficult. For instance, imagine building a predictive model that uses historical sales data to predict future sales. If the sales data only comes from one geographic location, the model may not generalize well to other regions.
- **Data Cleaning and Preparation:** Once we have acquired the data, we need to clean it and prepare it for analysis. This may involve removing errors, correcting inconsistencies, and transforming the data into a format that can be used by the model. Data cleaning and preparation is a critical step in the process of building predictive models. The quality of the data used to train a model has a direct impact on the accuracy of the model. The challenges in data cleaning and preparation while building predictive models can arise from a variety of sources. Some of the main challenges include dealing with data entry errors, missing or incomplete data, identifying and handling outliers, addressing data quality, consistency issues, and managing data imbalances. In addition, ensuring that the data is in the correct format and that any necessary transformations are performed correctly can be challenging, as well as making decisions about which features to include or exclude from the model. Moreover, selecting the appropriate modeling technique and performing appropriate model validation can also pose a challenge to data cleaning and preparation. Therefore, it is essential to establish a robust data cleaning and preparation framework that incorporates good data validation and data quality control practices, and also choosing appropriate features, modeling techniques, and validation metrics to ensure that the predictive model performs well and provides reliable insights [4].
- **Data Entry Errors:** Data entry errors are one of the most common problems with data quality. These errors can occur when data is entered into a system manually or when it is transferred from one system to another.
- **Missing Data:** Missing data can also be a problem for predictive modeling. This can occur when data is not collected for all of the records in a dataset or when data is lost or corrupted. Missing or incomplete data can arise for a variety of reasons, such as data entry errors, incomplete surveys, or participants dropping out of studies. Addressing missing data requires choosing an appropriate imputation method or selecting a sample with complete data, depending on the analysis goals and the amount of missing data.
- **Outliers:** Outliers are data points that are significantly different from the rest of the data. Outliers can skew the results of predictive models, so it is important to identify and deal with them. Outliers are observations that do not fit with the rest of the data and can distort the results of the analysis. Identifying outliers

requires visual inspection and statistical tests, such as box plots, scatterplots, or z-scores, and handling them can involve deleting them, transforming them, or creating a new category.

- **Data Imbalance:** Data imbalances can arise when the dataset is skewed toward one category or outcome, making it difficult to train predictive models. Addressing data imbalance requires selecting an appropriate sampling method, such as oversampling, under sampling, or synthetic minority over-sampling technique (SMOTE), or using different techniques for different classes.
- **Inconsistent Data:** Inconsistent data can also be a problem for predictive modeling. This can occur when the same data is stored in different ways or when different people enter the same data in different ways.
- **Data Bias:** Data bias occurs when the data is not representative of the population that you are trying to model. This can lead to inaccurate predictions. In addition to data cleaning, data preparation involves selecting and transforming the variables used to train the model. Decisions about which features to include or exclude from the model can be challenging, requiring domain knowledge, intuition, and empirical testing. Feature selection involves identifying the most relevant and informative variables that can predict the outcome of interest and reducing the number of variables to improve the model's performance and reduce overfitting. Last, selecting the appropriate modeling technique and performing appropriate model validation can also pose a challenge to data cleaning and preparation. The choice of modeling method depends on the analysis goals, the data characteristics, and the available resources and expertise.

This chapter contains four sections, Section 2.1 describes the predictive model and its method. Section 2.2 describes the literature survey of the predictive model. Section 2.3 describes the overcome technique to handle all challenges issues, and Section 2.4 describes the conclusion.

2.2 Literature Survey

Feature engineering is the process of transforming raw data into features that can be used to build predictive models. It is the process of selecting, transforming, and extracting useful features from the data set to improve the model's accuracy and minimizing overfitting. Feature engineering can be a laborious task since it involves identifying relevant variables that affect the outcome variable. In addition, feature engineering requires domain expertise, time consuming, intuition, and empirical testing to identify the most informative features, and it can be challenging in complex data sets. To overcome feature engineering challenges, we should leverage domain expertise to identify relevant features. We should also use data visualization techniques to identify correlations and patterns that can guide feature selection. In addition, they can use automated feature selection algorithms such as principal component analysis (PCA) and LDA [5].

2.2.1 Challenges in Feature Selection

It is the process of choosing the right features to use in a model. There are a number of different approaches to feature selection, each with its own advantages and disadvantages. Some of the most common feature selection methods include:

- **Filter Methods**: Filter methods select features based on their statistical properties, such as their correlation with the target variable or their variance.
- **Wrapper Methods**: Wrapper methods select features by iterative building and evaluating models on different subsets of features.
- **Embedded Methods**: Embedded methods select features as part of the model-building process.
- **Feature Creation**: It is the process of creating new features from existing features. There are a number of different ways to create new features, such as aggregation, transformation, and discretization. Some of the most common feature-creation methods include the following:
  - **Aggregation**: Aggregation is the process of combining multiple features into a single feature. For example, you could aggregate multiple sales data points into a single monthly sales figure.
  - **Transformation**: Transformation is the process of changing the scale or distribution of a feature. For example, you could normalize a feature to have a mean of 0 and a standard deviation of 1.
  - **Discretization**: Discretization is the process of converting a continuous feature into a categorical feature. For example, you could discretize a continuous age feature into age ranges, such as <18, 18–25, 25–35.
- **Feature Engineering for Different Models**: The same features may not be equally important for different models. For example, a linear regression model may be more sensitive to continuous features, while a decision tree model may be more sensitive to categorical features. Therefore, it is important to engineer features specifically for the model that will be used.

One of the main challenges in feature engineering is the selection of relevant features from the available data. Feature selection involves identifying the most relevant and informative variables that can predict the outcome of interest and discarding irrelevant or redundant variables. The process of feature selection can be challenging, as it requires assessing the importance of each feature, the degree of multicollinearity (correlation between variables), and the level of noise or irrelevant information in the dataset. Another challenge in feature engineering is the transformation of features to improve their predictive ability. Feature transformation involves converting data into a more useful form, such as normalizing distributions, scaling, and dimensionality reduction. Normalizing distributions involves transforming the data to follow a normal distribution, which is important for some machine learning algorithms that assume normality. Scaling involves standardizing the range of variables to ensure that they are comparable and have equal weight in the model. Dimensionality reduction involves reducing the number of features used in the model to improve its performance and reduce overfitting. Handling missing data is another challenge in feature engineering. Missing data can arise due to data entry errors, incomplete surveys, or participants dropping out of studies. Addressing missing data requires choosing an appropriate imputation method, such as mean imputation, regression imputation, or multiple imputation, depending on the analysis goals and the amount of missing data. Addressing outliers is another challenge in feature engineering. Outliers are observations that do not fit with the rest of the data and can distort the results of the analysis. Identifying outliers requires visual inspection and statistical tests, such as box plots, scatterplots, or z-scores, and handling them can involve deleting them, transforming them, or creating a new category [6]. Last, selecting the appropriate modeling technique and performing appropriate model validation can pose a challenge in feature engineering. The choice of modeling method depends on the analysis goals, the data characteristics, and the available resources and expertise. The use of proper model validation techniques, such as cross-validation, can help to ensure that the model is reliable and generalizable.

2.2.2 Challenges in Model Selection

There are various models, including linear regression, decision trees, random forests, and neural networks. Each model has its strengths and weaknesses. Therefore, selecting the appropriate model is essential for accurate predictions. The performance of the model depends on a number of factors, such as model complexity, model bias, model variance, data availability, and computational resources [7].

- **Model Complexity**: There are a variety of different models available, ranging from simple linear models to complex neural networks. The more complex the model, the more data it requires to train and the more likely it is to overfit the data.
- **Model Bias**: Model bias occurs when the model is not able to capture the true relationship between the features and the target variable. This can be caused by several factors, such as a lack of data, noise in the data, or a poorly chosen model.
- **Model Variance**: Model variance occurs when the model's predictions vary significantly from one data sample to another. This can be caused by a number of factors, such as a small sample size, noise in the data, or a poorly chosen model.

- **Data Availability**: The amount of data available can affect the model selection process. In general, more data is better, as it allows the model to learn the true relationship between the features and the target variable. However, collecting and storing large amounts of data can be expensive and time-consuming.
- **Computational Resources**: The computational resources required to train and evaluate models can vary significantly. For example, training a large neural network can require a powerful computer and a significant amount of time.

To overcome model selection challenges, data scientists should understand the business problem they want to solve and the data characteristics. They should also evaluate their models' performance using metrics such as root mean square error (RMSE), accuracy, precision, recall, and F1-score. In addition, they can use model selection techniques such as cross-validation and ensemble methods to improve model performance [8].

2.2.3 Challenges in Training and Evaluating the Model

Once you have chosen a model, you need to train it on your data. This involves feeding the data into the model and allowing it to learn the relationships between the features and the target variable. Once the model is trained, you need to evaluate its performance on a holdout dataset. Building accurate predictive models is a crucial task in various fields, including finance, healthcare, marketing, and engineering. These models are used to identify patterns, predict future trends, and make informed decisions. However, building a predictive model is challenging due to various factors, such as data quality, feature engineering, algorithm selection, training, and evaluation [9].

One of the primary challenges in training the model is the availability and quality of data. Inaccurate or incomplete data can affect the performance of the model, leading to biased or incorrect predictions. Furthermore, the data may contain outliers, missing values, or noisy features that can impact the training process. Data cleaning, preprocessing, and feature engineering are essential steps to ensure the quality and relevance of the data used to train the model. Another challenge in training the model is the selection of appropriate models, such as learning rate, regularization, or batch size. These hyperparameters control the behavior of the model during training and can affect its performance. Tuning the hyperparameters requires an iterative process of experimentation and evaluation. However, the number of hyperparameters and their possible combinations can be overwhelming, making the process of finding the optimal settings time-consuming and challenging [10].

The performance of the model heavily depends on the size and quality of the training dataset. A small dataset may not provide enough information to capture the complexity of the problem and may lead to overfitting, where the model memorizes the training data and fails to generalize to new data. In contrast, a large dataset can require extensive computational resources to train effectively, making it challenging to scale the training process. Furthermore, the size and distribution of the data can affect the performance of different algorithms and require different strategies for training. Evaluating the performance of the model is another significant challenge in building predictive models. Simply measuring accuracy or error rate may not provide sufficient insights into the model's behavior and may not generalize to new data. A robust evaluation process should include a variety of metrics, such as precision, recall, F1 score, or ROC curve, that capture different aspects of the model's performance. In addition, the evaluation process should include cross-validation, where the data is split into training and testing sets, and multiple models are evaluated on different splits to ensure the generalizability of the results [11].

2.2.4 Challenges in Deploying and Maintaining the Model Parameter Tuning

Once you have a trained model, you need to deploy it so that it can be used to make predictions. This may involve creating a web application, integrating the model into an existing system, or simply saving the model to a file. Once the model is deployed, you need to maintain it by monitoring its performance and retraining it as needed. Deploying and maintaining a predictive model is an important part of the machine learning process. However, it can be a challenging task, due to a number of factors [12].

Here are some of the challenges in deploying and maintaining models:

- **Model Complexity**: As the complexity of the predictive model increases, it may require additional hardware resources such as memory, processing power, or specialized computing infrastructure to support it. This can add to the cost of developing and deploying it. In addition, the more complex a model becomes, the greater the likelihood of failure in its deployment. This could happen because of an incorrect installation or configuration error, which makes deploying a working model even more problematic. For example, a large neural network may require a specialized environment to run in, and it may be difficult to update the model with new data.
- **Model Bias:** Identifying and correcting bias within a predictive model can be difficult and time-consuming. Model developers may need to review the training data to detect any data inaccuracies or model assumptions. Moreover, standard algorithms may require modification to eliminate inherent biases.

  The process of minimizing bias in a model may require adjustments to data inputs, algorithmic configurations, and model design. It may also require collaboration from domain experts to provide regular guidance and feedback to ensure that the model is fair and unbiased in its predictions.

- **Model Drift:** Model drift occurs when the distribution of data changes and differs significantly from the original data that was used to train the model. This can arise due to several reasons, such as changes in business or market conditions, new customer profiles, and user behavior patterns [13]. As a result, the model may make incorrect predictions or provide suboptimal results, which can result in a loss of accuracy and trust in the model's capabilities.

  Detecting model drift may be challenging as the changes could be subtle and occur over long durations. Model developers must always monitor models' performance in production to ensure they continue to produce reliable predictions. They may need to utilize tools and processes that enable them to collect data in real time and track changes to the model's output to detect any signals of potential drift.
- **Data Staleness:** Data staleness happens when the data used to train the model becomes outdated to the point that it no longer represents the current real-world scenario. This could be due to various reasons, such as changes in business conditions, market dynamics, or customer behaviors that have not been taken into account in the training data. When trained on such stale data, the model may fail to capture new patterns in the input data, which can lead to weakened predictive performance. Using stale data may lead to incorrect decisions and a lack of trust in the machine learning model. This problem can be particularly challenging in dynamic problem domains where data is continually evolving. In addition, detecting data staleness may be difficult, and it may require suitable techniques that involve regular updates and testing that ensure the model's inputs remain relevant and its predictive performance remains stable. To mitigate against data staleness, machine learning models must be regularly updated with fresh data that is representative of the current state of the world. This may require revisiting the model's features, checking their relevance, and training the model with updated data to reflect the latest trends and changing patterns. This process should be repeated regularly to ensure that the machine learning model remains accurate and able to provide reliable predictions over time [14].
- **Model Versioning**: Keeping track of different versions of a predictive model is essential as machine learning models are usually trained on data that changes over time, and a change in the training data can result in different model versions with varying performance. Also, the models are periodically updated with new features, new algorithms are tested, and the model's hyperparameters may be fine-tuned to improve its efficiency. Tracking these modifications provides an accurate way to understand how different changes have impacted the model's performance and the decision-making process. Moreover, tracking and monitoring the different versions of the machine learning model becomes particularly crucial when the model is deployed in a production environment, and changes

could result in feedback loops with significant effects. In such cases, monitoring the model's performance across different versions can assist in detecting discrepancies or abnormalities in the model's output. It enables developers to quickly remediate potential issues before they lead to disastrous results.

Thus, maintaining different versions of a predictive model can help create an audit trail of refined models, algorithms, and experiment results that provide valuable insights into the changes that have impacted the model's performance and its effectiveness over time. Tracking different versions of a model also aids in debugging, explaining, and troubleshooting decisions taken by the machine learning model [15].

- **Model Documentation**: Documentation of the machine learning model, such as its features, parameters, and performance metrics, can provide a complete summary of the model-meaning what it does and how well it does it. This documentation helps in tracking and replicating the model training and development over time and in comparing and contrasting different model versions. Also, documentation serves as an audit trail, providing a comprehensive record of model development and training. This information is valuable for troubleshooting, explaining, debugging, and reusing the model. In addition, keeping track of the model's features can empower data scientists to experiment with and reinterpret model behavior in novel and alternative ways. Detailed documentation of the model can help in comparing different models based on benchmarks and allow data scientists to describe the relative performance of the model accurately. Documentation of performance metrics provides insights into model errors and accuracy, training and validation loss, data drift, and regularization techniques. Finally, a well-documented machine learning model can enable communication among different stakeholders, including developers, engineers, and business users. A documented machine learning model can act as a means of knowledge exchange and transfer between teams, promoting transparency, accountability, and collaboration. Thus, documentation can help to ensure that the model is used correctly and that it can be easily updated and maintained [16].

Deploying and maintaining the model in a real-world scenario poses a new set of challenges that need to be overcome to ensure that the model performs as expected. One of the critical challenges in deploying and maintaining the model is parameter tuning. In this chapter, we will explore this challenge in depth and discuss possible solutions. Parameter tuning is the process of finding the best values for the parameters of the model. These parameters control the behavior of the model and determine its performance. In a real-world scenario, the data used by the model may not be the same as the data used to train the model. Furthermore, the environment in which the model operates may change over time, resulting in a shift in the distribution of the data. As a result, the model's parameters may need to be adjusted to adapt to these changes. Failure to do so may lead to a decrease in the performance and accuracy of the model. One common approach to parameter tuning is manual tuning, where the values of the parameters are adjusted by experts based on their experience and intuition. While this approach can be effective, it is time-consuming, expensive, and subjective. Moreover, manual tuning may not be scalable, especially when dealing with multiple models or large datasets. Another approach is automatic parameter tuning, where the model's parameters are adjusted using an optimization algorithm. This approach has proven to be effective, especially when dealing with complex models that have many parameters. However, automatic parameter tuning can also be costly, as it requires extensive computational resources, and may require additional data and time to perform the optimization [17]. Deploying and maintaining the model requires a robust monitoring and evaluation system that can detect performance changes and alert the user. This system should be designed to handle different types of errors and should be able to adapt to changes in the data distribution. Moreover, the system should be able to trigger re-training and parameter tuning when necessary, ensuring that the model stays up to date and reflects the current state of the environment.

2.2.5 Challenges in Interpretability, Explain Ability and Scalability

Building predictive models is an exciting task that involves multiple steps, from data preprocessing to feature engineering, model selection, and training. This process is increasingly becoming complex, as models are becoming more accurate and powerful, and businesses are demanding more from the technology. However, as models become more complex, a new challenge arises, which is the interpretability, explainability, and scalability of the models. Interpretability and explainability in predictive models refer to the ability to understand the model's inner workings and how it arrives at its predictions. In other words, it is the ability to demonstrate that the model is making its decision based on valid logic and not random factors [18]. Scalability is the ability of a model to handle large amounts of data. This is important because the amount of data that is available for training models is constantly growing. One of the critical challenges in achieving interpretability and explainability is the complexity of the models. Machine learning models can be very complex, requiring numerous factors to be taken into account when making predictions. As a result, explaining the decision-making process of those models can be challenging [19].

Another challenge is scalability. As data volumes continue to grow, models must be designed to handle large datasets, leading to increased complexity and reduced interpretability. It becomes difficult to explain how a particular decision was made when the model is processing millions of data points across multiple parameters.

Interpretability is the ability to understand why a model makes a particular prediction. This is important for a number of reasons, including [20]:

- **Trust**: Users need to trust that a model is making accurate predictions for it to be used in a decision-making process.
- **Fairness**: Models that are not interpretable can be more likely to discriminate against certain groups of people.
- **Continuous Learning**: If a model is not interpretable, it can be difficult to learn from its mistakes and improve its performance over time.

Explainability is a related concept to interpretability. Explainability is the ability to provide a human-understandable explanation for a model's prediction. This can be done in a number of ways, such as providing a list of the features that were most important in making the prediction or providing a graph that shows how the features interact to influence the prediction.

There are a number of challenges that need to be considered when building interpretable, explainable, and scalable models [21].

2.2.6 Challenges for Interpretability

- **Complex Models**: When we say a model is complex, it means that it contains a large number of parameters, layers, and nonlinear functions that make it difficult to understand how the model is making predictions. As a result, it becomes challenging to interpret the model's behavior and understand why it is making certain predictions, which can be a significant challenge in areas where interpretability is crucial, such as healthcare or finance. In addition, the complexity of these models makes it challenging to identify potential biases and errors, which can lead to detrimental effects on the model's performance. Overall, the complexity of machine learning models is one of the most significant challenges facing the field of AI, and researchers are actively working on developing techniques to improve the interpretability of these models.
- **Data Scarcity**: When we try to train a machine learning model that is interpretable, we need to have enough data for the model to learn meaningful patterns. However, obtaining a sufficient amount of data to train a model and still maintain interpretability can be challenging. This is because, in some cases, data may not be readily available or the cost of collecting additional data may be prohibitive.

  Moreover, in some domains, sensitive data (such as confidential medical records or financial information) may be subject to strict regulations, which can make collecting and sharing data more challenging. In addition, in cases where interpretability is required, the model's complexity may need to be constrained, which can limit the types of features that can be used, and can reduce the model's data requirements.

Obtaining enough data to train an interpretable model can be challenging, and often requires a careful balance between collecting enough data to enable meaningful learning, while also respecting privacy concerns and maintaining interpretability.

- **Cost**: When we develop interpretable models, it requires a considerable amount of time and effort from highly skilled professionals, including domain experts and data scientists. This includes tasks such as selecting and refining features, defining decision boundaries, creating visualizations, and optimizing model performance.

  Interpretable models may also require specialized hardware or software, such as graphics processing units (GPUs) or dedicated model interpretable development and deployment tools. These tools can be costly to acquire and maintain and may require additional training to use.

  Finally, deploying interpretable models in a production environment requires careful planning and coordination with other stakeholders in the organization. This includes ensuring that the model complies with relevant regulations and standards, providing documentation to support model decisions, and training end-users on how to use the model. Creating interpretable models can be an expensive process that requires significant expertise, specialized tools, and careful planning and coordination. However, the potential benefits of interpretable models, including improved accountability and trust in the decision-making process, make the investment worthwhile for many organizations [22].

2.2.6.1 Challenges for Explainability

- **Lack of Standards**: Machine learning models can be quite complex, and it is important for users to be able to interpret how the model arrives at its decisions. However, there are currently no established standards on how to explain the output of machine learning models. This makes a difficulty for users to understand and trust the model's output.

  Different machine learning models have different ways of explaining their predictions, and the quality and comprehensiveness of these explanations can vary widely. Furthermore, the explanation itself may be too complex to be easily understood by the user, particularly for deep learning models that have many layers with millions of parameters.

  These challenges can create obstacles to the adoption of machine learning in some domains, such as healthcare, finance, and other fields, where model interpretability is significant. Developing a standard for explaining machine learning models' predictions could enhance users' understanding and trust in those models, which could further encourage their adoption.

- **Complexity**: Machine learning models may draw on many factors to generate a particular output, including input data, model structure and parameters, and external factors. This complexity can make it difficult for users to understand how the output was determined and how to interpret the results. Providing explanations for machine learning models can be a time-consuming process, as it requires defining the model's decision rules and developing visualizations to support the interpretation. In addition, the potential number of factors that come into play when a model is generating an output can make the explanation quite extensive.

  Interpretability will depend on the complexity and structure of the machine learning model. Simpler models, such as linear regression or decision trees, can be more interpretable and explainable than more complex models like deep neural networks. Despite these challenges, explaining machine learning models' predictions is crucially important for enabling users to understand how these models work and gain insight into how decisions are made.

2.2.6.2 Challenges for Scalability

- **Computational Resources**: The process of training algorithmic models on large datasets usually requires substantial computational resources, as it involves processing immense amounts of data and calculating complex mathematical equations. This can include using powerful GPUs, which offer high-speed parallel processing capabilities, to speed up the model's training process. Deploying models can also demand significant computational resources. Large models may require large amounts of memory for storing the model, in addition to the processing power needed to finally make predictions. This can lead to added hardware costs and additional complexity in the deployment process.

  Overall, training and deploying large machine learning models can be computationally intensive and may require investment in specialized hardware or cloud services to support the model's processing demands. Moreover, this can be challenging for smaller organizations or those without significant computing resources, making it harder for them to adopt these models.
- **Data Storage:** When there is a need to store a large volume of data, it can take up significant disk space or memory. There are different types of data storage that can be used based on the size, structure, or type of data saved. Traditional storage methods like hard drives often have limited storage capacity and can be prone to failure. Enterprises are now turning toward cloud storage, an online storage service that can scale to accommodate larger volumes of data and provide more significant redundancy options.

  Moreover, the sizeable investment required for data storage can also include infrastructure, hardware, software, and data centers to monitor the data storage system, backup and recover the data when necessary, and provide security to protect the stored data.

Storing large amounts of data can require a significant investment in resources and infrastructure to ensure that it can be appropriately saved, managed, and secured.

A possible solution to the challenges of interpretability and explainability is the use of transparent models. These models are simpler, using fewer parameters, and are easier to understand. Decision trees and regression models are good examples of transparent models that can be used to build models with good interpretability and explainability. These models are also easier to scale, making them ideal for large datasets. However, the use of transparent models may come at a cost. As these models are simpler, they may not be able to capture the complexity of the real-world scenario and may result in less accurate predictions [23]. In some cases, the trade-off is making a choice between accuracy and interpretability, depending on the specific needs of the business.

Another solution to the challenges of interpretability and explainability is the use of visualizations and dashboards that communicate the model's output in a way that is easy to understand for the end-user. Visualizations can help in identifying patterns and correlations that are not easily revealed by the model's output [17].

2.2.6.3 Challenges in Ethical and Regulatory Challenges

Building predictive models has revolutionized the way we make decisions in various fields. From healthcare to finance to marketing, predictive models have been used to predict outcomes and guide decision-making. However, there are several ethical and regulatory challenges that need to be considered when building and using these models [24].

- **Fairness**: One of the biggest challenges is ensuring that predictive models are fair. This means that they should not discriminate against any particular group of people. Models can be biased if they are trained on data that is not representative of the population they are intended to serve. This can result in the model systematically disadvantaging certain groups or individuals. For example, a facial recognition model trained on mostly white faces can perform poorly when attempting to recognize people of color. This can result in individuals being falsely accused or even arrested due to the inability of the model to recognize their faces.
- **Privacy**: Another important ethical issue is privacy. Predictive models often rely on large amounts of data about individuals, and it is important to protect this data from unauthorized access. Models can contain sensitive data, such as health records or financial information, which must be protected to prevent unauthorized access. The use of predictive models in fields such as healthcare or finance can also raise ethical concerns about the use of personal information to inform decision-making.

- To address these concerns, organizations must adopt strict data privacy policies, including guidelines for the collection, storage, and use of data. It is also crucial to implement secure data storage and transmission systems to prevent breaches and unauthorized access.
- **Transparency:** It is also important to be transparent about how predictive models work. This means that users should be able to understand how the model was built and how it makes predictions. For example, a predictive model used to make medical decisions should be transparent about the factors that it considers when making a diagnosis.
- **Regulation**: Regulatory challenges while building predictive models can also arise due to the potential consequences of their use. For example, healthcare providers may use predictive models to guide treatment decisions, which can have significant implications for patient outcomes. As a result, regulatory bodies may require that the accuracy and reliability of these models be formally evaluated before their use in the clinical setting.

To address regulatory challenges, organizations must follow ethical guidelines and best practices, including transparency, accountability, and rigorous validation processes. This can involve conducting independent assessments of the model's accuracy and fairness, providing detailed documentation on how the model works, and implementing clear guidelines for its use. For example, the European Union's General Data Protection Regulation (GDPR) sets out a number of requirements for organizations that process personal data.

2.3 Few Suggestions to Overcome the Above Challenges

Building accurate predictive models requires a lot of work, but it can be a very rewarding experience. By overcoming the challenges of data cleaning and preparation, you can build models that can help you make better decisions and improve your business [24].

Here are some tips for overcoming the challenges of feature engineering for predictive modeling:

- **Clear Understanding of the Business Needs**: What are you trying to predict? What data do you need to make the prediction?
- **Identify the Data Sources**: Where is the data located? What format is the data in?
- **Explore the Data**: Get a good understanding of the data, including the distribution of the data, the relationships between the features, and the presence of any outliers.

- **Choose the Right Features**: Use a variety of techniques to select the right features for your model such as statistical tests, domain knowledge, and machine learning algorithms.
- **Create New Features**: Create new features from existing features using a variety of techniques such as aggregation, transformation, and discretization.
- **Engineer Features for Different Models**: Engineer features specifically for the model that will be used.
- **Prevent Overfitting**: Use regularization and cross-validation to prevent overfitting.
- **Deal with Data Sparsity**: Use techniques such as dimensionality reduction and feature selection to deal with data sparsity.
- **Be Patient and Persistent**: Feature engineering can be challenging, but it is an essential step in the machine learning process. By being patient and persistent, you can build models to help you make better decisions and improve your business.
- **Using Fair Machine Learning Algorithms**: There are a number of machine learning algorithms that have been designed to be fair. These algorithms can be used to build predictive models that are less likely to discriminate against any particular group of people.
- **Protecting Privacy**: There are a number of techniques that can be used to protect privacy when building and using predictive models. These techniques include anonymizing data, using differential privacy, and using secure multi-party computation.
- **Being Transparent**: It is important to be transparent about how predictive models work. This can be done by documenting the model, providing explanations for its predictions, and making the model available for inspection.
- **Complying with Regulations**: In some cases, it is necessary to comply with regulations when building and using predictive models. This can be done by working with a lawyer or consultant who specializes in data privacy and security.
- **Choose the Right Model**: Some models are more interpretable than others. For example, linear regression models are more interpretable than deep learning models.
- **Use Interpretability Techniques**: There are a number of techniques that can be used to make models more interpretable. These techniques include feature selection, dimensionality reduction, and visualization.
- **Document the Model**: It is important to document the model, including its features, parameters, and performance metrics. This documentation can help users to understand the model and to trust its predictions
- **Develop Standards**: There is a need to develop standards for how to explain a model's predictions. This would make it easier for users to understand and trust the model.

- **Automate Explanation**: There is a need to develop techniques for automatically explaining a model's predictions. This would reduce the cost and complexity of explaining models.
- **Use Cloud Computing**: Cloud computing can provide the computational resources needed to train and deploy large models.
- **Use Distributed Computing**: Distributed computing can be used to train and deploy models on multiple computers.
- **Use Data Compression**: Data compression can be used to reduce the amount of data that needs to be stored and processed.

2.4 Conclusion and Future Directions

Building predictive models presents many challenges, including ethical, privacy, and regulatory concerns. Addressing these challenges requires collective efforts from all stakeholders, including the development teams, policymakers, and users. Actions such as diverse data collection, privacy policies, and fair assessment processes can help mitigate the risks of bias and increase the trustworthiness of predictive models. The use of predictive models is expected to continue growing, with breakthroughs in technology, such as AI and machine learning, leading to even greater insights and innovation. However, it is important that we continue to prioritize ethical considerations and regulations to ensure that these models are used responsibly and effectively.

References

1 Li, Y., Yu, X., and Koudas, N. (2021). Data acquisition for improving machine learning models. *arXiv preprint arXiv:2105.14107.*

2 Yang, Y., Xia, X., Lo, D. et al. (2022). Predictive models in software engineering: challenges and opportunities. *ACM Transactions on Software Engineering and Methodology (TOSEM)* 31 (3): 1–72.

3 Hazen, B.T., Boone, C.A., Ezell, J.D., and Jones-Farmer, L.A. (2014). Data quality for data science, predictive analytics, and big data in supply chain management: an introduction to the problem and suggestions for research and applications. *International Journal of Production Economics* 154: 72–80.

4 McCue, C. (2014). *Data mining and predictive analysis: intelligence gathering and crime analysis*. Butterworth-Heinemann.

5 Li, R., Zhao, Z., Zhou, X. et al. (2014). The prediction analysis of cellular radio access network traffic: from entropy theory to networking practice. *IEEE Communications Magazine* 52 (6): 234–240.

6 Hofmanninger, J., Prayer, F., Pan, J. et al. (2020). Automatic lung segmentation in routine imaging is primarily a data diversity problem, not a methodology problem. *European Radiology Experimental* 4 (1): 1–13.
7 Krishnan, S., Haas, D., Franklin, M.J., and Wu, E. (2016). Towards reliable interactive data cleaning: a user survey and recommendations. *SIGMOD/PODS'16: International Conference on Management of Data*, San Francisco, CA (26 June–1 July 2016).
8 Zhang, S., Zhang, C., and Yang, Q. (2003). Data preparation for data mining. *Applied Artificial Intelligence* 17 (5-6): 375–381.
9 Bhatia, K. and Syal, R. (2017). Predictive analysis using hybrid clustering in diabetes diagnosis. In: *2017 Recent Developments in Control, Automation & Power Engineering (RDCAPE)*, Noida, India, 447–452. IEEE https://doi.org/10.1109/RDCAPE.2017.8358313.
10 Khurana, U., Samulowitz, H., and Turaga, D. (2018). Feature engineering for predictive modeling using reinforcement learning. *Proceedings of the AAAI Conference on Artificial Intelligence* (Vol. 32, No. 1), New Orleans, Lousiana, USA (2–7 February 2018).
11 Mao, J. and Jain, A.K. (1995). Artificial neural networks for feature extraction and multivariate data projection. *IEEE Transactions on Neural Networks* 6 (2): 296–317.
12 Wang, L., Wang, X., and Feng, J. (2006). On image matrix based feature extraction algorithms. *IEEE Transactions on Systems, Man, and Cybernetics, Part B (Cybernetics)* 36 (1): 194–197.
13 Ando, T. (2011). Predictive Bayesian model selection. *American Journal of Mathematical and Management Sciences* 31 (1-2): 13–38.
14 Paxton, C., Niculescu-Mizil, A., and Saria, S. (2013). Developing predictive models using electronic medical records: challenges and pitfalls. *AMIA Annual Symposium Proceedings* 2013: 1109. American Medical Informatics Association.
15 Paleyes, A., Urma, R.G., and Lawrence, N.D. (2022). Challenges in deploying machine learning: a survey of case studies. *ACM Computing Surveys* 55 (6): 1–29.
16 Ogumerem, G.S. and Pistikopoulos, E.N. (2019). Parametric optimization and control toward the design of a smart metal hydride refueling system. *AIChE Journal* 65 (10): e16680.
17 Nebeker, C., Harlow, J., Espinoza Giacinto, R. et al. (2017). Ethical and regulatory challenges of research using pervasive sensing and other emerging technologies: IRB perspectives. *AJOB Empirical Bioethics* 8 (4): 266–276.
18 Gade, K., Geyik, S.C., Kenthapadi, K. et al. (2019). Explainable AI in industry. *Proceedings of the 25th ACM SIGKDD International Conference on Knowledge Discovery & Data Mining*, pp. 3203–3204, Anchorage, AK (25 July 2019). New York: Association for Computing Machinery.

19 Palmer, A. and Greenhough, B. (2021). Out of the laboratory, into the field: perspectives on social, ethical and regulatory challenges in UK wildlife research. *Philosophical Transactions of the Royal Society B* 376 (1831): 20200226.

20 Zhao, L., Dong, Q., Luo, C. et al. (2021). DeepOmix: a scalable and interpretable multi-omics deep learning framework and application in cancer survival analysis. *Computational and Structural Biotechnology Journal* 19: 2719–2725.

21 Tahmassebi, A., Motamedi, M., Alavi, A.H., and Gandomi, A.H. (2022). An explainable prediction framework for engineering problems: case studies in reinforced concrete members modeling. *Engineering Computations* 39 (2): 609–626.

22 Ahmad, M.A., Eckert, C., and Teredesai, A. (2018). Interpretable machine learning in healthcare. *Proceedings of the 2018 9th ACM International Conference on Bioinformatics, Computational Biology and Health Informatics*, Washington DC USA (29 August 2018–1 September 2018).

23 Cohen, I.G., Amarasingham, R., Shah, A. et al. (2014). The legal and ethical concerns that arise from using complex predictive analytics in health care. *Health Affairs* 33 (7): 1139–1147.

24 Coeckelbergh, M. (2019). Artificial intelligence: some ethical issues and regulatory challenges. *Technology and Regulation* 2019: 31–34.

3

AI-driven Digital Twin and Resource Optimization in Industry 4.0 Ecosystem

Pankaj Bhambri[1], *Sita Rani*[2], *and Alex Khang*[3]

[1] *Department of Information Technology, Guru Nanak Dev Engineering College, Ludhiana, Punjab, India*
[2] *Department of Computer Science and Engineering, Guru Nanak Dev Engineering College, Ludhiana, Punjab, India*
[3] *Department of Artificial Intelligence, Global Research Institute of Technology and Engineering, Fort Raleigh City, NC, USA*

3.1 Introduction

In the era of Industry 4.0, the fusion of digital technologies with the physical world has redefined the landscape of industrial operations. This transformative wave has brought forth a paradigm shift in the way businesses manage their resources, from manufacturing facilities to supply chains and beyond. At the heart of this revolution lies the concept of the "Digital Twin," a virtual representation of physical assets and processes. When coupled with the power of artificial intelligence (AI), these AI-driven Digital Twins become potent tools for realizing unprecedented levels of resource optimization and operational efficiency [1–4].

This chapter embarks on an insightful exploration into the realm of AI-driven Digital Twins and their indispensable role within the Industry 4.0 ecosystem. It delves deep into the intricacies of how AI technologies are leveraged to create digital replicas of real-world assets and systems, enabling organizations to monitor, analyze, and optimize their operations with newfound precision and foresight. From manufacturing plants and energy grids to healthcare facilities and logistics networks, the applications of AI-driven Digital Twins are far reaching, promising substantial gains in terms of reduced downtime, improved product quality, enhanced sustainability, and competitive advantage [5–9].

As we navigate through this chapter, we will unravel the core components that constitute AI-driven Digital Twins, from data acquisition techniques to advanced

Intelligent Techniques for Predictive Data Analytics, First Edition.
Edited by Neha Singh, Shilpi Birla, Mohd Dilshad Ansari, and Neeraj Kumar Shukla.

Published 2024 by John Wiley & Sons, Inc.

modeling and analytics. Real-world use cases and success stories will illuminate the tangible benefits of adopting this innovative approach, while we also shedlight on the challenges, including data security and integration hurdles, that organizations must address on their journey toward AI-driven resource optimization [10–12].

Moreover, we will peer into the horizon, envisioning a future where AI-driven Digital Twins are not just tools but strategic assets, reshaping industries and fostering sustainable growth. Through this exploration, we aim to provide a comprehensive understanding of the symbiotic relationship between AI, Digital Twins, and Industry 4.0, offering valuable insights for researchers, industry practitioners, and decision-makers seeking to harness the full potential of these technologies in their pursuit of operational excellence and resource optimization [13, 14].

3.1.1 Overview of Industry 4.0

Industry 4.0, often referred to as the fourth industrial revolution, represents a fundamental transformation in the way industries operate and manage their processes. It is characterized by the integration of digital technologies, automation, data analytics, and the Internet of Things (IoT) into various aspects of manufacturing and production. Key components of Industry 4.0 include the following [15–19].

- **Digitization**: Industry 4.0 involves the digitization of physical assets and processes, turning them into digital representations that can be monitored and controlled in real time. This includes the use of sensors, RFID tags, and other data-capturing technologies.
- **Connectivity**: The seamless connectivity of devices and systems is a hallmark of Industry 4.0. Through IoT platforms and high-speed networks, machines, products, and people can communicate and collaborate effectively.
- **Data Analytics**: The massive amounts of data generated in Industry 4.0 are leveraged through advanced analytics, including machine learning and AI. This data-driven approach enables predictive maintenance, quality optimization, and informed decision-making.
- **Automation**: Smart factories and processes are heavily automated, reducing the need for human intervention in routine tasks. Robots, autonomous vehicles, and cobots (collaborative robots) are commonly used.
- **Interoperability**: Industry 4.0 systems emphasize interoperability, enabling-different devices and software to work together seamlessly. This facilitates a more flexible and adaptable manufacturing environment.
- **Customization**: Mass customization becomes feasible with Industry 4.0, as production processes can be easily adapted to meet individual customer needs without compromising efficiency.

- **Resource Efficiency**: Through real-time monitoring and optimization, Industry 4.0 aims to enhance resource utilization, reduce waste, and promote sustainability.
- **Cybersecurity**: With increased connectivity, there is a heightened focus on cybersecurity to protect sensitive data and prevent cyberattacks.
- **Human–Machine Collaboration**: While automation is a key element, Industry 4.0 also emphasizes collaboration between humans and machines. Workers are often equipped with digital tools to enhance their decision-making and productivity.

Industry 4.0 has far-reaching implications for various sectors, including manufacturing, logistics, healthcare, and energy. It promises to revolutionize industries by improving efficiency, reducing costs, and opening up new opportunities for innovation. However, it also presents challenges, such as the need for upskilling the workforce and addressing cybersecurity concerns. Overall, Industry 4.0 represents a pivotal moment in the history of industrialization, with the potential to drive significant economic and technological advancements [20–22].

3.1.2 Role of AI-driven Digital Twin

The role of AI-driven Digital Twins is pivotal in bridging the physical and digital realms, revolutionizing industries by providing real-time insights, predictive capabilities, and optimization opportunities. These digital replicas of physical assets and processes, empowered by AI and data analytics, enable organizations to monitor, control, and optimize their operations with unprecedented precision. They facilitate real-time monitoring and control, allowing for the immediate detection of anomalies and deviations, which, in turn, enables proactive maintenance, reduces downtime, and enhances operational efficiency. Moreover, AI-driven Digital Twins excel in predictive analytics, leveraging historical and real-time data to forecast future behaviors and issues. This empowers organizations to make informed decisions, optimize resource utilization, and schedule maintenance with precision, ultimately leading to cost savings, extended asset lifespans, and improved sustainability. Furthermore, they facilitate simulation and testing, accelerating product development, enhancing quality, and reducing R&D costs. In addition, their role in personalization and customization enhances customer experiences and patient outcomes, driving competitiveness across various sectors. AI-driven Digital Twins are poised to play a transformative role in shaping the future of industries, unlocking new levels of innovation and efficiency [23–30].

AI-driven Digital Twins play a pivotal role in Industry 4.0 by creating virtual replicas of physical assets, processes, and systems. These digital counterparts leverage AI and data analytics to provide real-time insights, predictive capabilities, and optimization opportunities. Here are two real-time examples that illustrate the role of AI-driven Digital Twins in various industries [31–34]:

- **Manufacturing and Predictive Maintenance**: In the manufacturing sector, AI-driven Digital Twins are used to monitor and maintain complex machinery. For instance, consider a large industrial pump used in a water treatment facility. An AI-driven Digital Twin of this pump continuously collects data from sensors embedded in the physical pump such as temperature, pressure, and vibration levels. This data is then analyzed in real time by AI algorithms. It plays the roles below:
  - **Predictive Maintenance**: The AI-driven Digital Twin can predict when the pump is likely to fail based on anomalies or deviations in sensor data.
  - **Optimize Maintenance Schedules**: By scheduling maintenance, only when necessary, companies can reduce downtime and save on maintenance costs.
  - **Improved Efficiency**: Constant monitoring allows for real-time adjustments to pump operations, optimizing energy consumption and extending the equipment's lifespan.
- **Smart Healthcare and Patient Monitoring**: In healthcare, AI-driven Digital Twins are employed for remote patient monitoring and personalized treatment. Consider a patient with a chronic heart condition. An AI-driven Digital Twin of the patient is created by continuously collecting data from wearable sensors, electronic health records, and even genetic information. It plays the roles below:
  - **Real-time Monitoring**: The AI-driven Digital Twin tracks the patient's vital signs, medication adherence, and lifestyle choices.
  - **Predictive Health Insights**: By analyzing this data, AI algorithms can predict potential health issues or deteriorations in the patient's condition.
  - **Personalized Treatment**: Doctors can use the Digital Twin to tailor treatment plans and interventions to each patient's specific needs, improving outcomes.

In both examples, AI-driven Digital Twins serve as a bridge between the physical and digital worlds, enabling proactive decision-making, reducing downtime, and optimizing resource allocation. They are valuable tools in industries where real-time data and predictive insights are critical for operational efficiency and enhanced customer experiences. These applications demonstrate the transformative power of AI-driven Digital Twins across diverse sectors [35–38].

3.2 Digital Twin Technology

Digital Twin technology refers to the creation of a virtual replica or simulation of a physical object, system, process, or entity. This virtual representation, known as a "Digital Twin," is a detailed and real-time counterpart of its physical counterpart, and it is often used to monitor, analyze, predict, and optimize the performance of the physical asset or system. Digital Twin technology is particularly prevalent in industries such as manufacturing, healthcare, aerospace, and infrastructure management, where it offers valuable insights and benefits [39–45].

3.2.1 Definition and Concepts

Digital Twin technology is a cutting-edge concept that involves the creation of a virtual representation or replica, known as a "Digital Twin," of a physical object, system, process, or entity. This Digital Twin is a detailed and dynamic counterpart of its physical counterpart, continuously synchronized with real-time data from sensors and other data sources. It aims to simulate the physical asset or system's behavior, attributes, and interactions, providing a comprehensive and interactive digital model. By leveraging advanced technologies such as IoT connectivity, data analytics, and AI, Digital Twin technology enables real-time monitoring, analysis, prediction, and optimization of the physical asset or process. It finds applications across various industries, offering insights, efficiencies, and innovations that bridge the gap between the physical and digital worlds, ultimately enhancing performance, decision-making, and resource utilization [46–51].

3.2.2 Components of a Digital Twin

A Digital Twin is a complex virtual model of a physical asset, process, or system that replicates its real-world counterpart with a high degree of accuracy. To achieve this fidelity, Digital Twins are composed of several interconnected components that work together to capture, simulate, and analyze data. Here are the key components of a Digital Twin [52–56]:

- **Physical Entity or Object**: The core of a Digital Twin is the physical asset, process, or system that is being replicated. This could be anything from a manufacturing machine and an aircraft engine to a city's traffic management system or even an individual's health parameters [57, 58].
- **Sensor and Data Acquisition**: Sensors and data acquisition devices are used to collect real-time data from the physical entity. These sensors can measure various parameters such as temperature, pressure, vibration, location, and more, depending on the type of asset or system.
- **IoT Connectivity**: Digital Twins rely on the IoT to establish connectivity between the physical entity and the digital counterpart. IoT devices and networks enable the seamless transmission of data from the physical world to the virtual world.
- **Data Processing and Storage**: The collected data is processed and stored in databases or cloud platforms. This data can be vast and varied, requiring robust storage and processing capabilities to handle the volume and complexity.

- **Digital Model or Representation**: The heart of the Digital Twin is the digital model or representation of the physical entity. This model includes geometry, behavior, attributes, and any relevant parameters that accurately mimic the real-world object or system. It is constantly updated based on the incoming data.
- **Simulation and Analytics Engines**: Digital Twins often incorporate simulation engines and analytics tools. These engines use the digital model and real-time data to simulate various scenarios, predict behavior, and provide actionable insights. Machine learning and AI algorithms are frequently employed for data analysis.
- **Visualization Interface**: To make the Digital Twin accessible to users, there is usually a visualization interface. This can be a 3D rendering, a dashboard, or a user-friendly application that allows operators, engineers, or decision-makers to interact with and understand the Digital Twin's data and insights.
- **Control and Feedback Loop**: In some cases, Digital Twins have a control element that enables users to make real-time adjustments to the physical entity based on insights derived from the virtual model. This closed-loop control mechanism can optimize processes and prevent issues.
- **Communication and Collaboration**: Digital Twins can facilitate communication and collaboration among stakeholders. For instance, in smart cities, multiple entities can use a Digital Twin to coordinate urban planning and development.
- **Cybersecurity Measures**: Given the sensitivity of the data and potential vulnerabilities, robust cybersecurity measures are essential to protect the Digital Twin ecosystem from cyber threats and unauthorized access.

3.2.3 Real-Time Data Integration

Real-time data integration in Digital Twin technology is the continuous and instantaneous process of incorporating live data from physical assets or systems into their digital counterparts. This integration ensures that the virtual model or Digital Twin remains synchronized with the real-world object or process, reflecting its current state and behavior in real time. Through sensors and IoT devices, data is collected from the physical asset, transmitted securely to a central repository, processed, and then used to update the Digital Twin's digital model. This real-time synchronization empowers organizations to monitor, analyze, and make data-driven decisions with precision, offering insights into performance, predicting issues, and optimizing operations on the fly. Real-time data integration is at the heart of the transformative power of Digital Twin technology across various industries, enabling dynamic analysis, predictive capabilities, and proactive maintenance [59–63].

3.3 Industry 4.0 Ecosystem

The "Industry 4.0 ecosystem" refers to the integrated network of digital technologies, processes, and stakeholders that define the fourth industrial revolution. At its core, Industry 4.0 leverages advanced innovations such as the IoT, AI, automation, and data analytics to revolutionize manufacturing and industrial practices. This ecosystem fosters interconnected smart factories, where machines, systems, and data collaborate seamlessly to optimize operations, increase efficiency, and enhance decision-making. It extends beyond the factory floor to encompass supply chains, logistics, and even customer interactions, promoting real-time data-driven insights, sustainability, and innovation [64–67]. In essence, the Industry 4.0 ecosystem represents a transformative shift toward intelligent, agile, and digitally empowered industrial systems that redefine how products are made and services are delivered.

3.3.1 Components and Technologies

Components and Technologies in the Industry 4.0 Ecosystem are described below:

- **IoT**: IoT devices and sensors are the backbone of Industry 4.0, providing real-time data collection and connectivity for machines and systems. They enable the monitoring and control of physical assets and facilitate data exchange within the ecosystem.
- **AI and Machine Learning**: AI algorithms analyze vast datasets to derive insights, make predictions, and optimize processes. Machine learning models are employed for predictive maintenance, quality control, and decision support across various industries.
- **Big Data Analytics**: Big Data technologies process and analyze large volumes of data generated by IoT devices and other sources. This analysis leads to actionable insights, enabling data-driven decision-making and process optimization.
- **Cloud Computing**: Cloud platforms provide scalable storage and processing capabilities, allowing organizations to store, access, and analyze data from anywhere. Cloud services are integral to handling the massive data generated in Industry 4.0.
- **Edge Computing**: Edge devices and edge computing solutions process data closer to its source (i.e., at the "edge" of the network). This reduces latency, enables faster decision-making, and is critical for real-time applications in Industry 4.0.
- **Cybersecurity**: With increased connectivity, robust cybersecurity measures are essential to protect data and systems from cyber threats. Industry 4.0 places a strong emphasis on securing the entire ecosystem, from sensors to cloud services.

- **Augmented Reality (AR) and Virtual Reality (VR)**: AR and VR technologies enhance human-machine interaction by overlaying digital information onto the physical world. They are used for training, maintenance, and remote assistance in Industry 4.0.
- **Blockchain**: Blockchain technology provides secure and transparent data sharing and transactions among stakeholders in the ecosystem. It is valuable for ensuring trust and traceability in supply chains and transactions.
- **Autonomous Robots**: These robots, equipped with AI and sensors, perform tasks autonomously. They are used in manufacturing, logistics, and even healthcare for tasks like material handling and surgery.
- **Additive Manufacturing (3D Printing)**: 3D printing allows for rapid prototyping, customization, and on-demand production. It is used in manufacturing to create complex and unique components.
- **Smart Sensors**: Advanced sensors are embedded in machines and systems to monitor variables such astemperature, pressure, and vibration. They play a crucial role in real-time data collection and predictive maintenance.
- **5G Connectivity**: 5G networks offer high-speed, low-latency connectivity, enabling real-time communication between devices and systems. This is essential for applications requiring instantaneous data transfer.
- **Digital Twins**: Digital Twins are virtual replicas of physical assets and processes, providing real-time monitoring, analysis, and optimization. They are central to Industry 4.0 for creating a digital counterpart of the physical world.
- **Human–Machine Interfaces (HMIs)**: Intuitive user interfaces and HMIs allow humans to interact with machines and systems. They range from touchscreen displays to voice commands and gesture recognition.
- **Supply Chain Visibility Tools**: These tools enable end-to-end visibility and optimization of supply chains, allowing for better demand forecasting, inventory management, and logistics planning.

3.3.2 IoT Integration

In the Industry 4.0 ecosystem, "IoT integration" is a fundamental and transformative element that underscores the convergence of physical processes and digital technologies. IoT integration involves the strategic deployment of IoT devices and sensors within industrial environments to enable real-time data acquisition and seamless connectivity. These IoT devices, meticulously embedded in machinery, equipment, and even products, function as sensory extensions of the industrial ecosystem. They continuously collect a wealth of data, encompassing variables such as temperature, pressure, machine performance, and environmental conditions. This data is then transmitted across interconnected networks, incorporating both wired and wireless communication protocols, facilitating the exchange of information between devices, centralized control systems, and cloud-based platforms.

The significance of IoT integration within Industry 4.0 is manifold. It empowers organizations with real-time visibility into the operational status of their assets, allowing for remote monitoring and control. Beyond mere observation, IoT integration leverages advanced analytics and machine learning to transform raw data into actionable insights. Predictive maintenance, a hallmark application, relies on IoT-generated data to forecast equipment failures or maintenance requirements, reducing costly downtime and optimizing resource allocation. Quality control mechanisms are enhanced as IoT sensors monitor and flag deviations in manufacturing processes, enabling immediate corrective actions to maintain product quality standards.

IoT integration also drives energy efficiency initiatives, tracking energy consumption patterns and identifying opportunities for optimization, thus contributing to sustainability goals. In the supply chain, IoT devices play a pivotal role in enhancing visibilityand tracking the real-time location, condition, and transit history of products. This resultsin improved logistics planning, inventory management, and overall supply chain resilience. The interconnected nature of IoT in Industry 4.0 fosters efficient communication among various systems and stakeholders, ultimately facilitating more informed decision-making, greater operational efficiency, and increased competitiveness. In essence, IoT integration is the linchpin of Industry 4.0, ushering in a new era of data-driven, interconnected, and intelligent industrial processes.

3.3.3 Cyber–Physical Systems

Cyber–Physical systems (CPSs) stand asa central and transformative component within the Industry 4.0 ecosystem, symbolizing the seamless integration of the physical and digital realms in industrial processes. CPSs merge physical assets, such as machines and sensors, with digital technologies, including real-time data analytics and connectivity, to formintelligent and interconnected systems. These systems facilitate the monitoring, controlling, and optimizingof physical processes with an unprecedented level of precision and efficiency. CPSs play a pivotal role in Industry 4.0 by facilitating real-time data collection and analysis, enabling predictive maintenance, enhancing automation, and supporting agile decision-making. They empower industries to respond rapidly to changing conditions and customer demands, improve product quality, and reduce operational costs. Furthermore, CPSs underpin the development of smart factories, autonomous vehicles, and responsive supply chains, ushering in a new era of advanced manufacturing and industrial excellence characterized by flexibility, adaptability, and data-driven innovation [68–70].

3.4 AI in Digital Twins

AI plays a pivotal role in enhancing the capabilities of Digital Twins, making them smarter and more impactful. AI in Digital Twins involves the integration of machine learning, data analytics, and advanced algorithms into the virtual replicas of physical assets or processes. This infusion of AI empowers Digital Twins to go beyond mere data visualization and monitoring; they can now analyze massive datasets in real time, identify patterns, predict future behavior, and offer actionable insights. For instance, AI-driven Digital Twins can predict equipment failures in manufacturing, enabling proactive maintenance. In healthcare, thesecan simulate patient responses to different treatments, personalizing healthcare plans. The synergy of AI and Digital Twins is revolutionizing industries by optimizing operations, improving decision-making, and driving innovation through predictive analytics and real-time intelligence [71–73].

3.4.1 Machine Learning and AI Algorithms

Machine learning and AI algorithms formthe backbone of AI in Digital Twins, bestowing these virtual replicas with intelligence and analytical prowess. These algorithms enable Digital Twins not only to replicate physical assets or processes but also to analyze data, recognize patterns, and predict future behaviors. In industries such as manufacturing, AI algorithms can sift through vast datasets from sensors embedded in machinery to forecast maintenance needs, preventing costly downtime and optimizing operational efficiency. In healthcare, they can simulate patient responses to different treatments, facilitating personalized care plans. Machine learning and AI bring a dynamic and proactive dimension to Digital Twins, enabling them to provide real-time insights, predictive capabilities, and data-driven recommendations. Their role in AI-driven Digital Twins is central, revolutionizing industries by automating decision-making, improving resource allocation, and accelerating innovation through the power of data analytics.

3.4.2 Data Analytics for Predictive Maintenance

Data Analytics for predictive maintenanceplays a pivotal role in integrating AI within Digital Twins, especially in industrial settings. This application leverages historical and real-time data gathered by sensors and IoT devices integrated into physical machinery and assets. The Data Analytics component within the Digital Twin utilizes advanced algorithms, often based on machine learning and AI, to analyze this data. By identifying patterns, anomalies, and trends, it can forecastwhen equipment is likely to fail or in need of maintenance.

One of the key advantages of this approach is that it enables organizations to shift from reactive maintenance, where assets are fixed only when they break down, to proactive and predictive maintenance. With AI-driven Digital Twins, maintenance needs can be anticipated well in advance, allowing companies to schedule maintenance during periods of lower production or when it is most cost-effective. This approach minimizes unexpected downtime, extends the lifespan of machinery, and reduces maintenance costs. Furthermore, by continually learning from new data, these AI algorithms improve their predictive accuracy over time, making maintenance schedules even more efficient.

In essence, Data Analytics for Predictive Maintenance within AI-driven Digital Twins transforms the industrial landscape by enhancing asset reliability, optimizing resource allocation, and ultimately ensuring smoother and more cost-effective operations. This not only improves the bottom line for businesses but also contributes to sustainability goals by reducing unnecessary resource consumption and waste.

3.5 Resource Optimization

Resource optimization involves the dynamic management and allocation of resources such as energy, materials, and manpower to maximize efficiency and minimize waste in industrial processes. AI-driven Digital Twins play a central role by continuously collecting and analyzing real-time data from sensors and IoT devices, which enables organizations to gain precise insights into resource utilization. These insights enable data-driven decision-making, helping businesses identify areas where optimization is possible [74, 75].

For example, in manufacturing, AI-driven Digital Twins can monitor energy consumption, machine performance, and production rates to fine-tune operations and reduce energy costs. In smart buildings, they can regulateheating, cooling, and lighting systems based on occupancy and environmental conditions, resulting in significant energy savings. In addition, AI algorithms can optimize supply chain logistics by forecastingdemand fluctuations and dynamically adjusting inventory levels, ensuring efficient resource allocation.

3.5.1 Importance in Industry 4.0

Resource optimization not only enhances operational efficiency but also contributes to sustainability goals by reducing resource consumption and environmental impact. It plays a vital role in Industry 4.0, helping organizations thrive in an increasingly competitive and resource-conscious landscape while driving progress towarda more sustainable and efficient future.

Resource optimization holds immense importance in the context of Industry 4.0 as it aligns perfectly with the overarching goals of this transformative industrial paradigm. In an era where competitiveness hinges on efficiency, sustainability, and agile decision-making, resource optimization empowered by AI-driven Digital Twins is a linchpin. It enables organizations to extract maximum value from their resources, whether it is energy, raw materials, or labor, by leveraging real-time data analytics and predictive capabilities. This not only translates into reduced operational costs and enhanced profitability but also aligns with the growing emphasis on environmental responsibility and sustainability. In Industry 4.0, where connectivity, data, and automation reign supreme, resource optimization is a strategic imperative that drives productivity, resilience, and competitiveness, positioning organizations for long-term success in the digital age.

3.5.2 Cost Reduction Strategies

Cost reduction strategies take center stage as an essential component of resource optimization. The integrationof AI and Digital Twins equips organizations with powerful tools to identify inefficiencies and streamline operations across various industries. By continuously monitoring and analyzing real-time data, these technologies can pinpoint areas of underutilized resources or instances of wastage. For instance, in manufacturing, predictive maintenance powered by AI-driven Digital Twins can mitigate costly breakdowns and unplanned downtime, leading to substantial savings in repair and replacement costs. Similarly, in smart buildings, resource optimization can dynamically adjust energy consumption based on occupancy patterns, yielding significant reductions in utility bills [76]. Cost reduction strategies not only bolster the bottom line but also enhance overall competitiveness, a critical objective within the Industry 4.0 paradigm, where efficiency and sustainability are paramount.

3.5.3 Energy Efficiency

Energy efficiency serves as a cornerstone of resource optimization. Through the convergence of AI and Digital Twins, organizations can achieve unparalleled levels of energy optimization and sustainability. These technologies enable real-time monitoring and analysis of energy consumption patterns, allowing precise adjustments in usage. For instance, in manufacturing, AI-driven Digital Twins can orchestrate the operation of machinery and equipment to match production demands, thereby minimizing energy wastage during idle periods. In smart buildings, they can dynamically control heating, cooling, and lighting systems in response to occupancy, reducing energy costs while maintaining occupant comfort [77]. Energy efficiency, powered by AI-driven Digital Twins, not only contributes significantly to cost savings but also aligns seamlessly with environmental responsibility, a pivotal aspect of Industry 4.0's commitment to sustainable and resource-conscious practices.

3.6 AI-driven Resource Allocation

It refers to the process of intelligently and dynamically distributing resources, such as manpower, materials, equipment, or energy, within an organization or system using AI technologies. This approach involves leveraging AI algorithms, machine learning models, and data analytics to optimize the allocation of resources in realtime or near-realtime [78]. AI-driven resource allocation aims to maximize efficiency, reduce waste, and enhance overall productivity by ensuring that resources are allocated to tasks or processes based on current demand, priorities, and available data.

In a manufacturing facility, AI-driven resource allocation can optimize production schedules by considering factors such asmachine availability, workforce capacity, and material availability. In a smart grid system, AI can allocate electricity generation and distribution resources dynamically to meet fluctuating demand while minimizing energy loss. In healthcare, AI-driven resource allocation can optimize patient scheduling, efficientlyallocate medical staff, and allocate resources based on the severity of cases.

Overall, AI-driven resource allocation plays a critical role in industries and systems where efficient utilization of resources is essential, helping organizations achieve cost savings, enhance operational effectiveness, and respond quickly to changing conditions.

3.6.1 Resource Monitoring

Resource monitoring, in the context of AI-driven resource allocation, is a fundamental process that involves the continuous, real-time surveillance of critical assets and components within an organization or system. This surveillance is achieved through the deployment of various monitoring mechanisms, including sensors, IoT devices, and data collection tools, which gather essential data related to resource availability, utilization, and performance. Resource monitoring forms the bedrock of informed decision-making in AI-driven resource allocation. By continuously collecting and analyzing data, organizations gain valuable insights into the state of their resources, enabling them to make precise and data-driven decisions regarding resource allocation.

For instance, in manufacturing, resource monitoring can track machine uptime, energy consumption, and material usage, providing a holistic view of production operations. In the context of smart buildings, it can monitor occupancy patterns, temperature, and lighting conditions to optimize energy consumption and space utilization. The data obtained from resource monitoring is not only used to understand current resource usage but also to predict future requirementsand detect anomalies or inefficiencies in real time.

Resource monitoring empowers AI-driven resource allocation systems to dynamically adjust resource allocation based on the most up-to-date information, ensuring that resources are allocated efficiently and effectively. In addition, it plays a pivotal role in predictive maintenance, enabling organizations to schedule repairs and maintenance activities precisely when needed, thus minimizing downtime and reducing maintenance costs. Resource monitoring serves as a foundational element in AI-driven resource allocation, enabling organizations to harness the power of data to optimize their resource utilization, enhance operational efficiency, and respond proactively to changing demands and conditions. It represents a crucial step toward achieving the goals of Industry 4.0 by facilitating intelligent decision-making and resource optimization across various domains and industries.

3.6.2 Roles and Types of Optimization Algorithms

Optimization algorithms are a key component of AI-driven resource allocation. These algorithms are essential for making intelligent decisions regarding the allocation of resources to achieve specific objectives. Here, we delve into the role and various types of optimization algorithms commonly used in this context:

- **Linear Programming (LP) and Integer Linear Programming (ILP)**: LP and ILP algorithms are fundamental optimization techniques that address linear relationships among variables. In resource allocation, LP models can allocate resources, such as labor, materials, and machines to tasks or projects to maximize efficiency or minimize costs while adhering to constraints. ILP extends this approach to problems where decision variables must take integer values, making it suitable for discrete resource allocation problems.
- **Genetic Algorithms**: Genetic algorithms are inspired by natural selection and evolution. They can be utilized to discover optimal solutions for resource allocation problems by generating and evolving a population of potential solutions across multiple generations. Genetic algorithms are particularly useful for solving complex, nonlinear resource allocation problems with many variables.
- **Simulated Annealing**: Simulated annealing is a probabilistic optimization algorithm inspired by the annealing process in metallurgy. It explores the solution space by iteratively modifying resource allocations while accepting less favorable solutions with a decreasing probability. This approach is

effective for finding near-optimal solutions in large and complex resource allocation scenarios.

- **Particle Swarm Optimization (PSO)**: PSO is an optimization technique inspired by the social behavior of birds or fish in swarms. It can be applied to resource allocation problems by simulating the movement of particles in a multi-dimensional space, with each particle representing a potential solution. The algorithm iteratively refines resource allocations to converge toward an optimal or near-optimal solution.
- **Reinforcement Learning**: Reinforcement learning algorithms, a subset of machine learning, can be employed for resource allocation by training agents to make allocation decisions based on past experiences and feedback. In dynamic environments, reinforcement learning can adapt resource allocation strategies over time to optimize performance.
- **Heuristic Algorithms**: Heuristic algorithms are rule-based approaches that may not guarantee an optimal solution but can provide practical solutions to resource allocation problems in a reasonable amount of time. Examples include greedy algorithms, which make locally optimal choices at each step, and metaheuristic algorithms such as Tabu search and ant colony optimization.
- **Mixed-Integer Nonlinear Programming (MINLP)**: MINLP algorithms manageresource allocation problems with mixed types of variables combining both continuous and discrete decision variables. They are suitable for complex optimization scenarios where both continuous and integer decisions need to be made.

3.6.3 Optimization Algorithms for AI-Driven Resource Allocation

In manufacturing, AI-driven resource allocation brings about a paradigm shift by optimizing resource utilization and operational efficiency. This transformative approach finds application in a range of use cases, revolutionizing traditional manufacturing processes. For instance, predictive maintenance, powered by AI algorithms and resource allocation models, enables manufacturers to schedule machinery repairs precisely when needed, reducing downtime, and ensuring that resources, such as labor and spare parts, are allocated efficiently. Smart production scheduling, another use case, harnesses real-time data to allocate labor and machines dynamically, adapting to fluctuating production demands and minimizing idle time. Furthermore, energy-efficient resource allocation in manufacturing ensures that energy is allocated in real time to meet production needs while minimizing consumption during off-peak periods, contributing to cost savings and sustainability goals. In these and many other use cases, AI-driven resource allocation empowers manufacturers to optimize their operations, enhance competitiveness, and adapt to the dynamic demands of the Industry 4.0 era.

3.7 Challenges and Consideration

Implementing AI-driven Digital Twins and resource optimization within the Industry 4.0 ecosystem not only brings immense potential but also presents notable challenges and considerations. First, data privacy and security become paramount, as the integration of IoT devices and the sharing of sensitive data increase vulnerability to cyber threats. Organizations must invest in robust cybersecurity measures to safeguard critical information. Second, there is a need for skilled personnel who can manage, interpret, and act on the data generated by these systems. Talent shortages in fields like data science and AI may pose obstacles. In addition, the upfront costs of implementing these technologies can be substantial, requiring careful budgeting and resource allocation. Furthermore, interoperability issues may arise when integrating diverse systems and technologies, demanding standardized protocols and seamless communication between components. At last, ethical considerations concerning data usage, transparency, and accountability must be addressed to ensure that AI-driven systems act responsibly and ethically. Addressing these challenges while leveraging the opportunities presented by AI-driven Digital Twins and resource optimization is crucial for organizations aiming to thrive in the Industry 4.0 landscape.

3.8 Future Trends

The future of AI-driven Digital Twins and resource optimization within the Industry 4.0 ecosystem holds promising and transformative trends. First, the convergence of technologies, such as 5G and edge computing, will enable even more real-time data processing and decision-making, allowing organizations to achieve unparalleled levels of responsiveness and agility. Second, the continued advancement of AI and machine learning algorithms will result in increasingly sophisticated Digital Twins capable of predictive and prescriptive analytics, going beyond mere monitoring to offer actionable insights and autonomous decision-making. Moreover, the concept of "Self-Optimizing Digital Twins" is emerging, where these virtual replicas can adapt and optimize themselves without human intervention, further enhancing resource efficiency.

Interoperability standards and frameworks are expected to mature, enabling seamless integration of diverse systems and facilitating cross-industry collaboration. As sustainability gains prominence, AI-driven Digital Twins will play a pivotal role in achieving eco-friendly manufacturing and resource usage, aligning with global environmental goals. Moreover, the democratization of AI tools and platforms will empower smaller enterprises to harness the benefits of these technologies, fostering innovation and competitiveness across industries. In summary, the future trends in AI-driven Digital Twins and resource optimization promise a landscape where organizations can anticipate, adapt, and optimize operations with unprecedented precision, efficiency, and sustainability, setting the stage for the continued evolution of Industry 4.0.

3.9 Conclusion

AI-driven Digital Twin and Resource Optimization in the Industry 4.0 Ecosystem encapsulate a profound journey through the transformative landscape of modern industrial practices. We have explored the convergence of cutting-edge technologies, such as AI, IoT, and Digital Twins, and their pivotal role in reshaping how resources are allocated and managed within the Industry 4.0 paradigm. The chapter has highlighted the profound impact of AI-driven Digital Twins in enhancing efficiency, reducing costs, and contributing to sustainability across diverse sectors, from manufacturing to supply chains and smart infrastructure.

However, as we haveobserved, the implementation of such advanced systems is not without its challenges, including cybersecurity concerns, the need for skilled talent, and ethical considerations. Nonetheless, these challenges are outweighed by the immense potential for optimizing operations, boosting competitiveness, and achieving environmental sustainability. Looking to the future, we anticipate a trajectory characterized by ever-sophisticated Digital Twins, increasingly autonomous resource optimization, and seamless interoperability, all underpinned by the democratization of AI tools. This chapter serves as a guiding light for organizations seeking to navigate the Industry 4.0 landscape, offering insights into how AI-driven Digital Twins and resource optimization can steer them toward a future of unprecedented efficiency and innovation in the digital age.

References

1 Abrol, N., Shaifali, R., Rattan, M., and Bhambri, P. (2005). Implementation and performance evaluation of JPEG 2000 for medical images. *International Conference on Innovative Applicationsof Information Technology for Developing World*, Shanghai, China (17–18 January 2006). New York: IEEE.

2 Bai, C., Dallasega, P., Orzes, G., and Sarkis, J. (2020). Industry 4.0 technologies assessment: a sustainability perspective. *International Journal of Production Economics* 229: 107776.

3 Bathla, S., Bhambri, P., and Jindal, C. (2007). Wearablecomputers: smart era in computing. *National Conference on Advances In Computer Technology and Applications*, GGNIMT, Ludhiana, Punjab, India (1–2 May 2007).

4 Bathla, S., Jindal, C., and Bhambri, P. (2007). Impact of technology on societal living. *International Conference on Convergence and Competition*, PCTE, Ludhiana, Punjab, India (1–2 March 2007), p. 14.
5 Bhambri, P., Gupta, O.P., Hans, S., and Singh, R. (2011). Conceptual translation as a part of GENE expression. *International Conference on Advanced Computing and Communication Technologies*, APIIT, Panipat, Haryana, India (5 November 2011), pp. 506–508.
6 Bhambri, P. (2010). Anadaptive and resource efficient hand off in recovery state in geographic adhocnetworks. *International Conference on Engineering Innovations-A Fillip to Economic Development,* CGI, Fatehgarh Sahib, Punjab (18–20 February 2010).
7 Bhambri, P. and Hans, S. (2010). Evaluation of integrated development environments for embedded system design. *Apeejay Journal of Management and Technology* 5 (2): 138–146.
8 Bhambri, P., Singh, M., Suresh, H., and Singh, I. (2010). Data mining model for protein sequence alignment. *Advanced Computing and Communication Technologies (IEEE Sponsored)*, APIIT, Panipat, Haryana, India (30 October 2010).
9 Bhambri, P. and Thapar, V. (2010). Iris biometric – a review. *National Conference on Cellular and Molecular Medicine*. PG Dissertation. Punjab Technical University, Jalandhar, Punjab, India.
10 Bhambri, P. and Hans, S. (2009). Direct non iterative solution based neural network for image compression. *PIMT Journal of Research* 2 (2): 64–67.
11 Bhambri, P., Hans, S., and Singh, M. (2009). Inharmonic signal synthesis andanalysis. *Technia-International Journal of Computing Science and Communication Technologies* 1 (2): 199–201.
12 Bhambri, P. and Singh, M. (2009). *Data Mining Model for Protein Sequence Alignment*. Jalandhar: Punjab Technical University.
13 Bhambri, P. and Thapar, V. (2009). Power distribution challenges in VLSI: an introduction. Presented at the *International Conference on Downtrend Challenges in IT*, PCTE, Ludhiana, Punjab, India (22 May 2009), pp. 63.
14 Bhambri, P., Hans, S., and Singh, M. (2008). Bioinformatics –friendship between bits & genes. *International Conference on Advanced Computing & Communication Technologies,* APIIT, Panipat, Haryana, India (8–9 November 2008), pp. 62–65.
15 Bhambri, P., Hans, S., and Singh, M. (2008). Fractal image compression techniques: afast fractal image compression method. *International Conference on Advanced Computing and Communication Technologies*, APIIT, Panipat, Haryana, India (8–9 November 2008), pp. 6576–6660.
16 Bhambri, P. and Singh, M. (2008). Direct non iterative solution based neural network for image compression. *PCTE Journal of Computer Sciences* 5 (2): 1–4.
17 Bhambri, P. and Singh, M. (2008). Biometrics: face recognition using eigenfaces. *Apeejay Journal of Management and Technology* 3 (2): 160–164.

18 Bhambri, P. and Singh, M. (2008). Image transport protocol for JPEG image over loss prone congested networks. *PIMT Journal of Research* 1 (1): 55–61.

19 Bhambri, P. and Nischal, P. (2008). Emerging trends of intersectoral growth of India with special reference to the service sector. *PIMT Journal of Research* 1 (2): 10–18.

20 Bhambri, P. and Nischal, P. (2008). Emerging new economy in telecommunication sector of India. *International Conference on Business Challenges & Strategies in Emerging Global Scenario*, PCTE, Ludhiana, Punjab, India (30 May 2008), p. 26.

21 Bhambri, P., Singh, H., and Gupta, S. (2007). Generation of fuzzy rules and membership function from training examples. *National Conference on Advances in Computer Technology and Applications,* GGNIMT, Ludhiana, Punjab, India (25 May 2007).

22 Bhambri, P., Nischal, P., and Gupta, S. (2007). Bioinformatics and computational biology. *National Conference on Some Aspects of Recent Trends in Engineering and Technology*, PCET, Lalru, Mohali (28 April 2007), p. 29.

23 Bhambri, P. and Mangat, A.S. (2005). Wireless security. *Abstract Book of UGC Sponsored National Conference on Emerging Computing Technologies*, HRMMV, Jalandhar, p. 101.

24 Bhambri, P. and Gupta, S. (2007). Interactive voice recognition system. *National Conference on Advancements in Modeling and Simulation*, LLRIET, Moga (20–21 January 2006), p. 107.

25 Bhambri, P. and Singh, M. (2006). Artificial intelligence. *National Seminar on E-Governance – Pathway to Progress*, SSIET, Derabassi, p. 14.

26 Bhambri, P. and Sharma, N. (2005). Priorities for sustainable civilization. *National Conference on Technical Education in Globalized Environment-Knowledge, Technology & The Teacher* (25–26 February 2005), SBBSIET, Padhiana, Jalandhar, p. 108.

27 Bhambri, P. and Gupta, S. (2005). A survey & comparison of permutation possibility of fault tolerant multistage interconnection networks. *National Conference on Application of Mathematics in Engineering and Technology* (26–28 March 2005), MIMIT, Malout, p. 13.

28 Bhambri, P. and Bhandari, A. (2005). Different protocols for wireless security. *National Conference on Advancements in Modeling and Simulation* (20–21 January 2006), LLRIET, Moga, p. 8.

29 Bhambri, P., Singh, I., and Gupta, S. (2005). Robotics systems. *UGC Sponsored National Conference on Emerging Computing Technologies* (29–30 September 2005), HRMMV, Jalandhar, p. 27.

30 Bhambri, P. and Singh, I. (2005). Electrical actuation systems. *National Conference on Application of Mathematics in Engineering and Technology*, MIMIT, Malout, p. 107.

31 Bhambri, P. and Mangat, A.S. (2005). Wireless security. *Abstract Book of UGC Sponsored National Conference on Emerging Computing Technologies*, HRMMV, Jalandhar, Punjab, India (29–30 September 2005), p. 101.

32 Bhambri, P., Sood, G., and Verma, A. (2005). Robotics design: major considerations. *UGC Sponsored National Conference on Emerging Computing Technologies*, HRMMV, Jalandhar, Punjab, India (29–30 September 2005), p. 100.

33 Bhambri, P., Singh, I., and Singh, J. (2005). Role of programming and mathematics in the design of robotics. *National Conference on Application of Mathematics in Engineering and Technology*, MIMIT, Malout (29–30 September 2005), p. 101.

34 Cañas, H., Mula, J., Díaz-Madroñero, M., and Campuzano-Bolarín, F. (2021). Implementing Industry 4.0 principles. *Computers & Industrial Engineering* 158: 107379.

35 Chauhan, M. and Rani, S. (2021). Covid-19: arevolution in the field of education in India. In: *Learning How to Learn Using Multimedia* (ed. D. Mishra and Y. Chuang), 23–42. Springer.

36 Chen, B., Wan, J., Shu, L. et al. (2017). Smart factory of Industry 4.0: key technologies, application case, and challenges. *IEEE Access* 6: 6505–6519.

37 Contreras, J.D., Garcia, J.I., and Pastrana, J.D. (2017). Developing of Industry 4.0 applications. *International Journal of Online Engineering* 13 (10): 30.

38 Chopra, S. and Bhambri, P. (2011). *A New Method of Edge Detection in Mammographic Images*. Jalandhar: Punjab Technical University.

39 Dhanalakshmi, R., Vijayaraghavan, N., Sivaraman, A.K., and Rani, S. (2022). Epidemic awareness spreading in smart cities using the artificial neural network. In: *AI-Centric Smart City Ecosystems* (ed. A. Khang, S. Rani, and A.K. Sivaraman), 187–207. CRC Press.

40 Garg, D. and Bhambri, P. (2011). A novel approach for fusion of multimodality medical images. *CiiT International Journal of Digital Image Processing* 3 (10): 576–580. http://www.ciitresearch.org/dl/index.php/dip/article/view/DIP062011001.

41 Garg, D. and Bhambri, P. (2011). *A Novel Approach for Fusion of Multi-Modality Medical Images*. Jalandhar: Punjab Technical University.

42 Ghobakhloo, M. (2020). Industry 4.0, digitization, and opportunities for sustainability. *Journal of Cleaner Production* 252: 119869.

43 Goel, R. and Gupta, P. (2020). Robotics and industry 4.0. In: *A Roadmap to Industry 4.0: Smart Production, Sharp Business and Sustainable Development* (ed. A. Nayyar and A. Kumar), 157–169. Springer.

44 Grewal, H.K. and Bhambri, P. (2006). Globe-IT: globalization of learning through open based education and information technology. *International Conference on Brand India: Issues, Challenges and Opportunities*, LIM, Phagwara, Punjab, India (29–30 April 2006), p. 24.

45 Gupta, S., Nischal, P., and Bhambri, P. (2007). Multimodal biometrics: enhancing security level of biometric system. *National Conference on Emerging Trends in Communication & IT*, RIEIT, Nawanshahr (24–25 February 2007), pp. 78–81.

46 Gupta, S., Nischal, P., and Bhambri, P. (2007). Mathematical modeling of EMG (ELECTROMYOGRAM). *International Conference on Global Trends in IT*, PCTE, Ludhiana (31 March 2007), p. 14.

47 Gupta, S., Nischal, P., and Bhambri, P. (2007). DNA computing: an emerging trend. *National Conference on Emerging Trends in Communication & IT*, RIEIT, Nawanshahr (24–25 February 2007), p. 267.

48 Gupta, S., Kaur, R., and Bhambri, P. (2007). Common channel signaling system #7: a global standard for telecommunications. *National Conference on Emerging Trends in Communication & IT*, RIEIT, Nawanshahr (24–25 February 2007), pp. 130–134.

49 Gupta, S., Nischal, P., and Bhambri, P. (2007). Data encryption standard (DES) algorithm with Diffie-Hellman key exchange. *National Conference on Emerging Trends in Communication & IT*, RIEIT, Nawanshahr (24–25 February 2007), pp. 135–143.

50 Garg, D. and Bhambri, P. (2006). A competitive market is pushing site search technology to new plateaus. *International Conference on Brand India: Issues, Challenges and Opportunities*, LIM, Phagwara (29–30 April 2006), p. 34.

51 Habib, M.K. and Chimsom, C. (2019). Industry 4.0: sustainability and design principles. *Presented at the 2019 20th International Conference on Research and Education in Mechatronics (REM).*

52 Jain, V.K. and Bhambri, P. (2005). *Fundamentals of Information Technology & Computer Programming*, fourth reprint. Katson Publishers. ISBN 81-88458-59-7.

53 Kamra, A. and Bhambri, P. (2007). *Computer Peripherals & Interfaces*. Pune: Technical Publications: Pune. ISBN 978-81-8431-308-6.

54 Kaur, G., Bhambri, P., and Sohal, A.K. (2006). Review analysis of economic load dispatch. *Proceedings of National Conference on Future Trends in Information Technology*, SJPMLIET, Radaur, Yamuna Nagar, Haryana, India (6–7 March 2005).

55 Kothandaraman, D., Manickam, M., Balasundaram, A. et al. (2022). Decentralized link failure prevention routing (DLFPR) algorithm for efficient internet of things. *Intelligent Automation and Soft Computing* 34 (1): 655–666.

56 Kumar, P., Banerjee, K., Singhal, N. et al. (2022). Verifiable, secure mobile agent migration in healthcare systems using a polynomial-based threshold secret sharing scheme with a blowfish algorithm. *Sensors* 22 (22): 8620.

57 Kumar, R., Rani, S., and Awadh, M.A. (2022). Exploring the application sphere of the internet of things in industry 4.0: a review, bibliometric and content analysis. *Sensors* 22 (11): 4276.

58 Machała, S., Chamier-Gliszczyński, N., and Królikowski, T. (2022). Application of AR/VR technology in Industry 4.0. *Procedia Computer Science* 207: 2990–2998.

59 Rani, S., Kataria, A., Kumar, S., and Tiwari, P. (2023). Federated learning for secure IoMT-applications in smart healthcare systems: a comprehensive review. *Knowledge-based Systems* 274: 110658.

60 Rani, S., Kataria, A., and Chauhan, M. (2022). Cyber security techniques, architectures, and design. In: *Holistic Approach to Quantum Cryptography in Cyber Security* (ed. S. Bhushan, M. Kumar, and P. Kumar), 41–66. CRC Press.

61 Rani, S., Kataria, A., Chauhan, M. et al. (2022). Security and privacy challenges in the deployment of cyber-physical systems in smart city applications: state-of-art work. *Materials Today: Proceedings* 62: 4671–4676.

62 Rani, S., Mishra, R.K., Usman, M. et al. (2021). Amalgamation of advanced technologies for sustainable development of smart city environment: a review. *IEEE Access* 9: 150060–150087.

63 Rattan, M., Bhambri, P., and Shaifali (2005). Information retrieval using soft computing techniques. *Presented at the National Conference on Bio-informatics Computing,* TIET, Patiala, Punjab, India (18–19 March 2005), pp. 58–60.

64 Rattan, M., Bhambri, P., and Shaifali (2005). Institution for a sustainable civilization: negotiating change in a technological culture. *Presented at the National Conference on Technical Education in Globalized Environment- Knowledge, Technology & The Teacher,* SBBSIET, Padhiana, Jalandhar (25–26 February 2005).

65 Salkin, C., Oner, M., Ustundag, A., and Cevikcan, E. (2018). A conceptual framework for Industry 4.0. In: *Industry 4.0: Managing the Digital Transformation* (ed. A. Ustundag and E. Cevikcan), 3–23. Springer.

66 Saucedo-Martínez, J.A., Pérez-Lara, M., Marmolejo-Saucedo, J.A. et al. (2018). Industry 4.0 framework for management and operations: a review. *Journal of Ambient Intelligence and Humanized Computing* 9: 789–801.

67 Singh, I., Salaria, D., and Bhambri, P. (2010). Comparative analysis of JAVA and AspectJ on the basis of various metrics. *International Conference on Advanced Computing and Communication Technologies (IEEE Sponsored),* APIIT, Panipat, Haryana, India (30 October 2010), pp. 714–720.

68 Singh, P. and Bhambri, P. (2009). Alternate organizational models for ports. *Apeejay Journal of Management and Technology* 2 (2): 9–17.

69 Singh, P., Bhambri, P., and Sohal, A.K. (2006). Security in local networks. *Proceedings of National Conference on Future Trends in Information Technology,* SJPMLIET, Radaur, Yamuna Nagar, Haryana, India (6–7 March 2005).

70 Singh, M., Singh, P., Kaur, K., and Bhambri, P. (2005). Database security. *Proceedings of National Conference on Future Trends in Information Technology,* SJPMLIET, Radaur, Yamuna Nagar, Haryana, India (6–7 March 2005), pp. 35–51.

71 Singh, M., Bhambri, P., and Kaur, K. (2005). Network security. *Proceedings of National Conference on Future Trends in Information Technology*, SJPMLIET, Radaur, Yamuna Nagar, Haryana, India (6–7 March 2005), pp. 155–161.

72 Singh, P., Singh, M., and Bhambri, P. (2005). Internet security. *Abstract Book of U.G.C. Sponsored National Seminar on Network Security and Its Implementations*, Doaba College, Jalandhar, Punjab (1 March 2005), p. 22.

73 Singh, P., Singh, M., and Bhambri, P. (2005). Security in virtual private networks. *Abstract Book of U.G.C. Sponsored National Seminar on Network Security and Its Implementations*, Doaba College, Jalandhar, Punjab, India (1 March 2005), p. 11.

74 Singh, P., Singh, M., and Bhambri, P. (2005). Embedded systems. *U.G.C. Sponsored National Seminar on Embedded Systems,* KMV, Jalandhar (24–25 January 2005), pp. 10–15.

75 Singh, P., Singh, M., Bhambri, P., and Sachdeva, R. (2005). A comparative study: security algorithms. *U.G.C. Sponsored National Seminar on Network Security and Its Implementations*, Doaba College, Jalandhar, Punjab, India (1 March 2005), p. 26.

76 Singh, P., Singh, M., and Bhambri, P. (2004). Interoperability: aproblem of component reusability. *Presented at the International Conference on Emerging Technologies in IT Industry*, PCTE, Ludhiana, Punjab, India (27 November 2004), p. 60.

77 Thapar, V. and Bhambri, P. (2009). Context free language induction by evolution of deterministic pushdown automata using genetic programming. *Presented at the International Conference on Downtrend Challenges in IT*, PCTE, Ludhiana, Punjab, India (22 May 2009), p. 33.

78 Yadav, V.S., Singh, A., Raut, R.D. et al. (2022). Exploring the application of Industry 4.0 technologies in the agricultural food supply chain: a systematic literature review. *Computers &Industrial Engineering* 169: 108304.

4

Predictive Analytics in Healthcare

N. Venkateswarulu[1,2], P. Pavan Kumar[1], and O. Obulesu[1]

[1] *Department of Computer Science and Engineering, Koneru Lakshmaiah Education Foundation, Hyderabad, Telangana, India*
[2] *Department of Computer Science and Engineering, G. Narayanamma Institute of Technology and Science, Hyderabad, Telangana, India*

4.1 Predictive Analytics

A subset of advanced analytics known as predictive analytics uses historical data along with statistical modeling, data mining, and machine learning to forecast future events. Utilizing trends in this data, businesses use predictive analytics to identify potential hazards and opportunities. Big data and data science are frequently associated with predictive analytics. Companies are currently flooded with data, which can range from log files to photos and videos and is stored in various data repositories within an organization. Data scientists employ deep learning and machine learning algorithms to detect patterns and forecast future events to obtain insights from this data. Neural networks, decision trees, and logistic and linear regression models are a few of these statistical methods [1]. These modeling methods build upon earlier prediction learnings to derive new predictive understandings. Models for predictive analytics are created to analyze historical data, identify patterns, track trends, and use that knowledge to forecast future trends. Classification, clustering, and time series models are standard predictive analytics models. In healthcare, predictive analytics is used to track many kinds of infections and identify and manage the care of people with chronic illnesses. To better understand the detection and management of such diseases, there is a need for health-mined medical information using predictive analytics.

Intelligent Techniques for Predictive Data Analytics, First Edition.
Edited by Neha Singh, Shilpi Birla, Mohd Dilshad Ansari, and Neeraj Kumar Shukla.

Published 2024 by John Wiley & Sons, Inc.

Based on the medical records of patients who had previously been given a diagnosis, many researchers have developed predictive models. Impressive predictions of patients with a high survival rate were made by these models [2]. To diagnose, monitor, or treat medical disorders, "medical imaging" refers to several technologies used to see the human body. Each technology type provides specific information regarding the part of the body being investigated or treated in relation to potential illness, damage, or the efficacy of medical care. The medical imaging, pharmaceuticals, and clinical research datasets give insights to correctly classify, cluster, and predict illnesses in their early stages. Standard supervised classification models, such as decision trees, random forest, neural networks, Naïve Bayes, and logistic regression do the segmentation of the data, the unsupervised clustering models, such as hierarchical, mean-shift, density-based spatial clustering of applications with noise (DBSCAN), expectation-maximization (EM) clustering using the Gaussian mixture models (GMMs), and k-means clustering do the grouping of similar data and time series models such as autoregressive (AR), moving average (MA), ARMA, and ARIMA (autoregressstion integrated moving average) models does the plotting of trends, seasonality, and cyclical behavior of the data.

The healthcare sector today generates enormous amounts of data from the documentation of data related to patients' health, data pertaining to medical devices, data related to clinical results, data from lab tests, data related to drug research, images with graphics, audio, and video data, data related to health insurance, data related to health policies, and data related to patients' assessments. This generated data is gathered collectively in both structured and unstructured form. These data must be digitized to be used in the current new media and information age [3]. The digitization of medical data will also aid in delivering care that is more effectively enhanced in quality, lowering the cost of medical facilities. Medical care federations can offer equipment and technology that are conveniently accessible because of the digitization of the information. This goes further in looking at the digital data, producing a vital understanding of the healing process. Big data analytics offers massive storage for all forms of data, higher processing capacity, and the ability to manage and handle essentially infinite simultaneous tasks or jobs when used with the Hadoop framework, an open-source software platform for data collection and storage. By improving operational procedures and, by extension, results, big data outperforms the prior deficient position in the medical care business. Big Data Analytics has enabled doctors to make quick judgments based on results. Businesses in the medical care sector are benefiting from the rise of analytics. By choosing the best analytical tools, gathering informative data, sharing data via electronic health records and electronic medical records, and exchanging medical information, healthcare data is reviewed. The development of the medical care industry has standardized, enhanced, and allowed for the early identification and diagnosis of many ailments, as well as the ability to treat them quickly.

4.2 Predictive Analysis in Medical Imaging

Medical imaging describes the methods and procedures used to produce images of a person's internal organs for use in clinical research and medical intervention [4]. Various medical disorders can be identified, tracked, and treated with the help of these photographs. Medical imaging modalities come in various forms, each with its own special applications and guiding principles.

A few of the most typical kinds of medical imaging are:

1) **X-ray**: Using electromagnetic radiation, X-ray imaging produces images of some soft tissues and bones. It frequently identifies dental issues, lung infections, and fractures.
2) **Computed Tomography (CT)**: To create cross-sectional images of the body, CT scans combine X-ray technology with computer processing. This gives precise details on the inside structures and helps make a variety of diagnoses, such as those for tumors, vascular problems, and internal traumas.
3) **Magnetic Resonance Imaging (MRI)**: Using a powerful magnetic field and radio waves, MRI creates precise images of the brain, organs, and soft tissues. It is particularly helpful for checking the joints, abdominal organs, and central nervous system.
4) **Positron Emission Tomography (PET)**: A radioactive tracer that generates positrons is injected during a PET scan. A PET scanner records the interaction between positrons and the tissue in the area, creating pictures that display metabolic activity. PET is beneficial for cancer staging and diagnosis.
5) **Nuclear Medicine**: In nuclear medicine, a small amount of radioactive substance is injected into the body and monitored by a specialized camera. This makes physiological processes visible and can be applied to the diagnosis of ailments such as thyroid disease, cancer, and bone abnormalities.
6) **Ultrasound**: High-frequency sound waves are used in ultrasound imaging to provide real-time pictures of the inside organs and tissues of the body. It is frequently employed throughout pregnancy to monitor the growing fetus and to see organs such as the heart, liver, and kidneys.
7) **Mammography**: Specialized X-ray images of the breast tissue are produced by mammograms. They are employed in the early identification and screening of breast cancer.
8) **Fluoroscopy**: A real-time X-ray imaging technology called fluoroscopy enables the visualization of moving components, including the circulatory system

and digestive system. In procedures like angiography and barium tests, it is frequently employed.

By offering noninvasive means to investigate the interior organs and bodily processes, these diverse imaging modalities have revolutionized medical practice. They help medical experts identify diseases accurately, devise effective treatments, and track the effectiveness of therapy.

Using data analysis methods, frequently powered by machine learning and artificial intelligence (AI), predictive analysis in medical imaging aims to forecast outcomes or offer insights from medical pictures [5]. This strategy may improve patient outcomes, diagnostic procedures, and healthcare resource utilization.

Predictive analysis is used in medical imaging in the following ways: *1. Treatment Response Monitoring, 2. Diagnosis and Risk Assessment, 3. Early Detection, 4. Disease Progression Prediction, 5. Personalized Treatment Planning, 6. Resource Optimization, 7. Population Health Management,* and *8. Reducing Radiologist Workload.*

1) **Treatment Response Monitoring**: In determining how patients are reacting to therapies, predictive analysis might be helpful. Algorithms can spot tiny changes that might not be immediately visible to the human eye by comparing medical photos obtained before and after therapy. This is especially helpful in the treatment of cancer, where it is vital to keep track of the tumor's size and treatment response.
2) **Diagnosis and Risk Assessment**: Large datasets of medical picture data can be used to train machine learning algorithms to identify patterns linked to specific disorders. These algorithms can help radiologists make precise and timely diagnoses once they have been trained. An algorithm may, for instance, examine lung X-rays to find patterns suggestive of lung disorders such as pneumonia or lung cancer.
3) **Early Detection**: Compared with conventional approaches, predictive analysis can help in spotting potential health risks earlier. Algorithms can identify small irregularities that can point to the early onset of a disease by examining photos and patient data.
4) **Disease Progression Prediction**: Predictive models can predict the course of some diseases by looking at longitudinal imaging data collected over time from patients. Clinicians can use this information to plan suitable interventions and therapies. Predictive analysis, for instance, may be able to foresee cognitive deterioration in neurodegenerative diseases like Alzheimer's disease based on changes in brain MRI images.
5) **Personalized Treatment Planning**: Predictive models can help with creating treatment plans that are specific to each patient's needs. An algorithm may, for instance, examine a patient's breast cancer biopsy images to forecast the

likelihood of response to various treatment options, assisting physicians in selecting the best course of action.

6) **Resource Optimization**: Healthcare organizations can distribute resources more effectively by using predictive analysis. For instance, hospitals can prioritize their treatment and resources by identifying which patients most likely require immediate intervention based on their imaging data.
7) **Population Health Management**: Understanding illness frequency and distribution can be aided by trend and pattern analysis across imaging data collected from a community. Strategies and initiatives for public health can be guided by this knowledge.
8) **Reducing Radiologist Workload**: Radiologists can benefit from AI-driven predictive algorithms that automatically recognize and flag problems in medical pictures. Radiologists are not replaced by this; rather, it allows them to concentrate on cases that call for their knowledge.

Predictive analysis in medical imaging necessitates high-quality data, carefully thought-out algorithms, and stringent validation to guarantee accuracy and dependability. To maintain patient safety and data privacy, regulatory and ethical issues are crucial when using AI-based systems in a healthcare context.

4.3 Predictive Analytics in the Pharmaceutical Industry

The research, development, production, and marketing of pharmaceutical treatments and remedies are the primary goals of the pharmaceutical industry, which is a subsector of the healthcare component. These medications are meant to be used in the management, prevention, and treatment of a variety of illnesses and ailments. By developing and manufacturing efficient and secure medicinal agents, the pharmaceutical sector significantly contributes to the improvement of public health.

Key aspects of the pharmaceutical industry include *Research and Development (R&D)*: This is the procedure for finding and creating new medications. It entails intensive scientific investigation, including the identification of possible therapeutic targets, the design and testing of molecules, preclinical research, and clinical trials to test safety and effectiveness in humans. *Clinical Trials*: Clinical trials are meticulous investigations carried out on people to judge the efficacy and safety of novel medications. They are carried out in stages involving controlled testing on progressively larger volunteer or patient populations. *Regulatory Approval*: Government health organizations like the European Medicines Agency (EMA) or the U.S. Food and Medicine Administration (FDA) must approve a medicine before it can be marketed and sold [6]. These organizations evaluate the safety and effectiveness of the medicine based on the scientific data from clinical trials. *Manufacturing*: Drugs must be produced by pharmaceutical companies in accordance with strict quality criteria. This entails mass production, testing for quality control, and ensuring that each batch is consistent. *Marketing and Sales*: Pharmaceutical firms launch marketing and sales campaigns to advertise their goods to consumers and healthcare professionals after medicine is licensed. This includes public relations efforts, advertisements, and communications with medical professionals. *Intellectual Property and Patents*: R&D expenditures are essential when creating a new medicine. Pharmaceutical companies frequently apply for patents to protect their inventions, which give them exclusive rights to manufacture and market the drug for a predetermined time period, usually 20 years. *Generic Drugs*: When a patent on a drug expires, other businesses can make generic versions of it. Generic medications are identical to brand-name medications in terms of active components, dosage form, and intended purpose but are frequently offered at a reduced price. *Biopharmaceuticals and Biotechnology*: Biopharmaceuticals, which are medicines made utilizing biotechnology methods, are now a part of the industry. Protein-based therapeutics, monoclonal antibodies, and gene therapies are some of these medications. *Drug Pricing and Access*: Drug accessibility and pricing have been the subject of discussion and worry because some novel therapies can be expensive and limit patient access. This has sparked conversations on the necessity of striking a balance between innovation and cost. *Global Impact*: With businesses operating abroad and working with researchers, healthcare providers, and regulatory organizations in several nations, the pharmaceutical sector has a global presence.

The pharmaceutical sector is distinguished by its dual purpose of raising money and enhancing human health. It faces issues with innovation, regulatory compliance, ethical issues, and market dynamics, much like any other industry. Delivering effective therapy to patients while juggling these difficulties remains a difficult task.

In the pharmaceutical sector, predictive analytics refers to the use of data analysis methods, frequently fueled by machine learning and AI, to forecast outcomes, trends, and patterns in a variety of areas related to medication research, production, marketing, and sales. Predictive analytics can assist pharmaceutical firms in making knowledgeable decisions, streamlining processes, and boosting overall productivity by utilizing historical and current data.

Predictive analytics is used in the pharmaceutical sector in the following ways:

Drug Discovery and Development: Through the examination of molecular architectures, biological information, and prior success rates, predictive analytics can help in the identification of possible medication candidates. By reducing the number of molecules that are most likely to have the desired pharmacological effects, helps hasten the drug discovery process.

Clinical Trial Optimization: By predicting patient enrolment rates, calculating trial completion times, and identifying potential difficulties, predictive models can aid in the optimization of clinical trial design. This makes studies run more effectively, lowering costs and speeding up the time it takes for new medications to reach the market.

Safety and Pharmacovigilance: Predictive analytics can be used to track unfavorable outcomes and identify potential safety issues with pharmaceuticals after they have been put on the market. Companies can see trends and patterns that might point to new safety issues by examining patient data and reports of adverse events.

Supply Chain Management [7]: Predictive analytics can help pharmaceutical companies optimize their supply chains. This entails anticipating drug demand, controlling inventory levels, and making sure that medicines are accessible where and when they are required.

Manufacturing Optimization: In manufacturing operations, predictive analytics can be used to foresee equipment breakdowns, quality discrepancies, and production inefficiencies. This assures consistent product quality and helps prevent disruptions.

Market Forecasting: To forecast consumer demand for particular medications or therapies, predictive models might examine market trends, patient demographics, and competitive environments. Marketing and sales strategies can be influenced by this information.

Personalized Medicine: By examining patient data, genetics, and illness features, predictive analytics can help in the development of individualized treatment programs. This enables pharmaceutical firms to customize medications for specific individuals, boosting treatment results.

Drug Pricing and Reimbursement: With the help of predictive analytics, the best price plans for pharmaceuticals may be developed, taking into account things like market demand, development expenses, and the competitive environment. This aids businesses in determining pricing that balances patient access and profitability.

Patient Adherence and Engagement: By examining previous patient data and behavior, predictive analytics can forecast patient adherence to medicine. Using this data, interventions can be created to increase patient compliance and outcomes.

Regulatory Compliance: By examining data pertaining to regulatory standards and norms, predictive models can help in anticipating potential compliance concerns. This enables pharmaceutical firms to proactively handle compliance difficulties.

Intellectual Property Management: By examining market patterns and competitive environments, predictive analytics can be used to evaluate the potential worth of intellectual property, including patents and innovations.

It is crucial to remember that applying predictive analytics in the pharmaceutical sector necessitates having access to high-quality data, cutting-edge analytical tools, and data science and domain knowledge expertise. When applying predictive analytics in this heavily regulated sector, it is crucial to take ethical issues, patient privacy, and regulatory compliance into account.

4.4 Predictive Analytics in Clinical Research

Clinical research is the systematic evaluation of novel therapies, medications, medical devices, and interventions in humans to assess their efficacy, safety, and effectiveness [6]. It is a crucial stage in the creation of medical innovations and is essential to enhancing patient care and expanding medical knowledge.

Clinical research involves several stages and follows a clear process:

Preclinical Research: Preclinical research is conducted on new medications or treatments prior to human testing. To evaluate the compound's safety profile, probable mechanisms of action, and preliminary effectiveness, laboratory and animal studies are required. Clinical trials can be started by researchers if the results are encouraging.

Clinical Trials: There are various phases of clinical trials, each with a distinct goal:

- **Phase 1**: These studies concentrate on determining the drug's safety, dosage, and potential side effects and involve a small number of healthy volunteers.
- **Phase 2**: To ascertain the drug's efficacy and further assess safety, a larger number of patients are studied during this phase.
- **Phase 3**: These studies, which involve a greater number of patients, are designed to validate the drug's effectiveness while also keeping track of its negative effects and evaluating it against other available therapies.
- **Phase 4**: Also referred to as post-marketing monitoring, these studies occur after medicine has received approval and is available on the market. They acquire more data regarding the advantages and risks of the drug while also keeping an eye on its long-term security.

Informed Consent: Individuals are given comprehensive information about the trial's goals, methods, potential risks and advantages, and their rights as participants prior to enrolling in a clinical trial. An important ethical need to guarantee that people make informed decisions is informed consent.

Randomized Controlled Trials (RCTs): RCTs, a popular kind of clinical trial, randomly assign participants to either the treatment group, which receives the intervention under study, or the control group, which receives a placebo or a conventional treatment. Randomization reduces bias and gives a starting point for comparing treatment results.

Endpoints: Specific endpoints are used in clinical trials to gauge results. These endpoints could include clinical results (such as a decrease in tumor size), surrogate markers (such as a decrease in blood pressure), or patient-reported outcomes (such as evaluations of quality of life).

Data Collection and Analysis: To evaluate the safety and effectiveness of the intervention, clinical trial data are gathered and analyzed. Conclusions on the effects of the treatment and the importance of discovered differences can be reached using statistical analysis.

Regulatory Approval: Researchers submit a new drug application (NDA) or an equivalent to regulatory organizations like the FDA in the United States or the EMA in Europe following the successful completion of clinical trials demonstrating safety and efficacy. To determine whether to approve the intervention for public use, these bodies examine the data.

Ethical Considerations: Strict ethical standards, including the defense of participants' rights and welfare, must be followed in clinical research. To guarantee that ethical standards are upheld, study protocols are reviewed and approved by Institutional Review Boards (IRBs) or Ethics Committees.

Post-Market Surveillance: Following a drug's or therapy's approval and release to the public, continuing monitoring keeps track of its long-term effectiveness and safety. Patient outcomes and adverse events are also continuously monitored.

Collaboration between researchers, healthcare providers, regulators, and patients is essential in the field of clinical research since it is dynamic and always changing. It plays a crucial role in expanding medical knowledge, enhancing patient care, and commercializing cutting-edge medicines.

Using data analysis methods, frequently powered by machine learning and AI, predictive analytics is used in clinical research to foresee and make well-informed judgments regarding many elements of clinical trials and research [7]. Predictive analytics can aid in the optimization of trial design, patient recruitment, data administration, and other crucial activities by analyzing historical and current data.

Predictive analytics is used in clinical research in the following ways:

Patient Recruitment and Enrolment: By examining patient data and electronic medical records, predictive analytics can help identify possible trial participants. Researchers can speed up recruitment and shorten trial durations by focusing on patient demographics that are likely to fit the trial criteria.

Trial Design Optimization: Clinical trial design can be made more effective and efficient with the help of predictive models. Researchers can estimate the best sample size, treatment arms, and outcomes to produce significant results by simulating various trial designs and circumstances.

Site Selection: Based on previous data about patient recruitment rates, data quality, and site performance, predictive analytics can assist find the top clinical trial sites. This guarantees that trials are carried out in locations with a higher chance of success.

Real-time Monitoring: Predictive models can examine real-time data from clinical trial locations to spot prospective problems such as unforeseen adverse events or protocol violations. This makes prompt intervention and corrective measures possible.

Patient Adherence and Dropout Prediction: The likelihood of a patient dropping out of the trial can be anticipated using predictive analytics. Researchers can address any problems and guarantee data integrity using this knowledge.

Data Quality and Validation: By comparing data from various sources, predictive models can assist in identifying problems with and discrepancies in data quality. By doing this, it is ensured that the data is reliable and correct.

Risk Assessment and Management: Clinical trial researchers can proactively address problems that could negatively affect trial success by using predictive analytics to detect potential risks and barriers.

Optimizing Clinical Trial Supply Chain: Based on enrolment rates, treatment schedules, and other variables, predictive models can foresee the need for medicine or therapy supplies. This makes it possible to guarantee that there is a sufficient supply of research goods.

Protocol Feasibility: By examining historical data and identifying potential issues with patient recruitment, outcomes, and other aspects, predictive analytics can evaluate the viability of a clinical trial plan.

Patient Stratification and Personalized Medicine: The identification of patient subgroups that are more likely to benefit from treatment can be aided by predictive models. This is especially important for research on personalized medical strategies.

Regulatory Compliance: By spotting potential problems or deviations, predictive analytics helps verify that trial protocols are in conformity with regulatory requirements.

Adaptive Trials: Adaptive trial designs, in which trial parameters are changed based on interim analyses, can be supported by predictive analytics. This enables researchers to maximize resource usage and experiment results.

Predictive analytics must be used in clinical research, which necessitates access to high-quality data, cutting-edge analytical tools, and knowledge of clinical trial techniques and data science. To ensure the ethical and efficient use of predictive analytics in clinical research, patient privacy, regulatory compliance, and other issues must be properly addressed.

4.5 AI for Disease Prediction

Utilizing various data-driven methodologies, illness prediction – also referred to as disease risk prediction or disease forecasting – estimates the probability that a person will contract a specific disease or medical condition [8]. To determine a person's risk of contracting a certain disease, this method is based on the analysis of various factors, including demographic data, medical history, lifestyle decisions, genetic markers, and environmental exposures. The goal of disease prediction is to make early intervention, individualized care, and preventive measures possible.

Illness forecasting functions in many ways, like *Data Collection*, Individuals' pertinent information is gathered, including demographics (age, gender, and ethnicity), medical histories (preexisting conditions, family history), lifestyle choices (diet, exercise, and smoking), genetics (genetic markers linked to disease risk), and environmental considerations (pollution exposure, location). *Feature Selection*, relevant features or variables are chosen from the acquired data for analysis. These characteristics may include biomarkers, medical history, genetic markers, and lifestyle elements associated with the condition of interest. *Data Preprocessing*, data is standardized, cleansed, and made ready for analysis. Outliers are found, missing values are resolved, and data is modified as necessary. *Model Development*, predictive models are created using statistical methods and machine learning algorithms. These models are developed utilizing historical data with known results (disease presence or absence). *Feature Engineering*, to improve the model's ability to predict outcomes, pertinent features are converted or integrated. Feature engineering may entail choosing subsets of already-existing variables or inventing new variables. *Model Training*, a subset of the data is used to train the predictive model. Depending on how complex the problem is, many methods can be utilized, including logistic regression, random forests, support vector machines, and neural networks. *Validation and Testing*, to evaluate the performance and generalizability of the trained model, it is tested on a different set of data (validation or test set). Accuracy, sensitivity, specificity, and area under the receiver operating characteristic curve (AUC–ROC) are common evaluation criteria. *Model Evaluation and Interpretation*, the model's effectiveness is evaluated, and the relative significance of the various elements is established. This aids in understanding the elements that influence illness risk the most. *Prediction*, the model can be used to forecast disease risk for new individuals based on their personal features and data once it has been evaluated and declared accurate. *Clinical Application*, healthcare professionals can use predictive models to help them identify patients who are more likely to contract a specific disease. This enables focused interventions, preventive actions, and individualized healthcare advice.

Continuous Improvement, as new data becomes available, predictive models can be modified, increasing their accuracy and usefulness.

It is important to remember that disease prediction algorithms cannot reliably predict when diseases will manifest. Based on the facts, they provide probabilities and risk assessments. Ethics, patient privacy, and data security are essential because disease prediction models use sensitive personal data. Before the models are appropriately and evidence-based integrated into clinical care, they must undergo extensive validation.

AI has shown immense potential in disease prediction by leveraging its ability to analyze large and complex datasets, identify patterns, and make accurate predictions. Here is how AI is used for disease prediction:

Data Collection and Integration: For accurate predictions, AI algorithms need large, varied datasets. This contains health history, genetic data, lifestyle information, and other pertinent data. To create a comprehensive picture of a person's health, AI can combine data from genomic databases, wearable technology, and electronic health records.

Feature Selection and Engineering: The most accurate features or variables from the data that predict the risk of disease are automatically found by AI algorithms. To improve the performance of the model, various features are combined and altered through feature engineering.

Machine Learning Algorithms: Different machine learning algorithms are used by AI to create predictive models. Logistic regression, random forests, support vector machines, gradient boosting, and neural networks are a few examples of typical algorithms. Each algorithm excels in identifying particular relationships and patterns in the data.

Deep Learning and Neural Networks: For complicated and high-dimensional data, deep learning techniques, particularly neural networks, are well suited. Recurrent neural networks (RNNs) are utilized for sequential data, such as time-series medical data, whereas convolutional neural networks (CNNs) are frequently employed for image-based illness prediction.

Model Training and Validation: The performance of AI models is validated using different datasets after they have been trained on historical data to determine how well they generalize to fresh data. Techniques for cross-validation aid in ensuring the model's dependability.

Predictive Accuracy: By detecting minute patterns and interactions that might not be visible to human observers, AI models can accurately forecast the risk of contracting diseases. Thus, early identification and intervention are made possible.

Personalized Risk Assessment: By examining a person's particular combination of characteristics, such as genetic predisposition, lifestyle decisions, and medical history, AI enables personalized risk assessment.

Real-time Monitoring: Using wearables and remote sensors, AI-powered systems can continuously monitor a person's health in real time. Any alterations that could portend a disease risk can be quickly identified.

Early Detection and Prevention: Even before clinical signs appear, AI can assist in identifying those who are more likely to contract particular diseases. Early intervention and preventive measures are made possible by this.

Integration into Clinical Practice: AI-based illness prediction models can be included in clinical processes and electronic medical records. These technologies can help healthcare professionals make informed decisions and suggest individualized interventions.

Challenges and Considerations: The requirement for high-quality data, algorithm interpretability, ethical considerations, and regulatory approval are only a few of the obstacles that AI for disease prediction must overcome despite its promise. It is essential to ensure the models are transparent, comprehensible, and bias-free.

Diverse Applications: Numerous ailments, including cancer, cardiovascular disease, diabetes, infectious infections, and neurological disorders, are treated with AI. For early identification, risk classification, predicting treatment outcomes, and outcome prognosis.

By facilitating more proactive, individualized, and efficient patient care, AI-driven disease prediction has the potential to revolutionize healthcare [9]. Realizing its potential in clinical practice, however, requires meticulous validation, responsible application, and tight coordination between physicians, data scientists, and regulatory organizations.

4.6 Medical Image Classification for Disease Prediction

Utilizing machine learning and deep learning techniques, medical image classification for illness prediction entails analyzing medical images and classifying them according to the presence or absence of particular diseases or conditions. Early detection, diagnosis, and treatment planning can all be aided by this strategy

Anticipating diseases, medical images are classified into:

1) **Data Collection:** Patients' medical images, including X-rays, MRIs, CT scans, mammograms, and histopathological slides, are gathered. The categorization model uses these photos as input data.
2) **Data Preprocessing:** Preprocessing is done on raw medical images to improve their quality and prepare them for analysis. Resizing, normalization, noise reduction, and contrast enhancement are examples of preprocessing techniques.

3) **Data Labeling:** Medical professionals assign each image a sickness or condition that corresponds to it. To build a strong classification model, accurate labeling is necessary.
4) **Feature Extraction:** Traditional machine learning algorithms take the relevant information from the photos and extract them. These characteristics could include edge information, texture, shape, and intensity.
5) **Deep Learning and Convolutional Neural Networks (CNNs):** CNNs, in particular, have demonstrated astounding performance in the classification of medical images. In particular, complex visual patterns benefit from CNNs' ability to automatically learn hierarchical features from the images.
6) **Model Training:** Based on a labeled dataset, the classification model is trained. The model gains the ability to identify patterns and characteristics that distinguish between various disease categories throughout training.
7) **Validation and Hyperparameter Tuning:** To evaluate the trained model's performance, a different dataset is used to validate it. To get the best outcomes, hyperparameters like learning rates and network design are tweaked.
8) **Performance Evaluation:** Metrics such as accuracy, sensitivity, specificity, recall, and F1-score are used to assess classification performance. These metrics show how well the algorithm can categorize positive and negative cases.
9) **Data Augmentation:** Techniques for artificially boosting the size of the training dataset are frequently utilized. To produce variations, this entails applying adjustments to the photos, such as rotation, flipping, and cropping.
10) **Transfer Learning:** Transfer learning entails optimizing pretrained deep learning models (like ImageNet) for medical image classification tasks from a starting point of pretrained deep learning models. This method makes use of the information gained from huge image databases.
11) **Ensemble Methods:** Multiple models are combined using ensemble methods to enhance classification performance overall. The robustness and generalizability of the model can be improved by using strategies like bagging and boosting.
12) **Clinical Integration:** After being trained and validated, the classification model may be included in clinical processes to help radiologists and other healthcare practitioners evaluate medical images and come to reliable diagnoses.
13) **Challenges and Considerations:** The necessity for annotated data, model interpretability, correcting the class imbalance, and managing various image modalities are all difficulties in medical image classification. Additionally crucial are ethical considerations and regulatory compliance.

By offering quicker and more accurate diagnoses, medical picture classification for disease prediction has the potential to revolutionize medical diagnosis [10] (Table 4.1).

Table 4.1 List of the latest Predictive Analytics Tool's. Adapted from [13].

Predictive analytics tool	Description
IBM SPSS Modeler	With the help of this comprehensive predictive analytics tool, users may create predictive models without any programming knowledge. For data pretreatment, modeling, evaluation, and deployment, it provides a visual interface.
Microsoft Azure Machine Learning	This application, which is a component of Microsoft's Azure cloud services, enables users to create, use, and administer machine learning models. It integrates with other Azure services and offers a number of algorithms.
KNIME	Users can construct data pipelines for predictive analytics and data science using the open-source platform KNIME. It provides a visual and modular workflow construction technique.
SAS Predictive Analytics	Users can create and apply prediction models using the toolset that SAS provides for predictive analytics. For advanced analytics, it contains a variety of statistical and machine learning methods.
Alteryx	Although its main strengths are in data blending and preprocessing, Alteryx also provides predictive analytics capabilities. It features a drag-and-drop interface that enables users to create prediction models.
RapidMiner	An open-source tool for predictive analytics called RapidMiner offers a visual user interface for prepping data, modeling it, and deploying it. Numerous machine learning algorithms are supported.
Google Cloud AI Platform	Machine learning model development, training, and deployment tools are available through Google's AI platform. TensorFlow and Google Cloud services are integrated.
Python and R	These are programming languages that are frequently used for predictive analytics, not tools per se. Predictive modeling makes considerable use of libraries like scikit-learn, XGBoost, LightGBM, and TensorFlow (for Python) and caret, randomForest, and xgboost (for R).
H2O.ai	An open-source platform called H_2O offers predictive analytics and machine learning capabilities. It can be used for applications such as classification, regression, and clustering and supports a variety of techniques.
DataRobot	The goal of the automated machine learning platform DataRobot is to streamline the creation and use of prediction models. It automates a number of machine learning pipeline phases.

To achieve its successful integration into clinical practice, it calls for meticulous validation, domain knowledge, and cooperation between medical professionals and data scientists.

There exists many significant number of well-liked predictive analytics solutions being used by organizations and data professionals as of most recent knowledge updated in recent days [11, 12], the following resources were helpful.

Predictive analytics is a dynamic field thus new tools may have been developed since my previous update. It is crucial to evaluate the precise requirements of your projects and select tools that are in line with the objectives and domain knowledge of your company. To make wise selections, always look at the most recent facts.

References

1 Jain, D., Kadecha, B. and Iyer, S. (2019). A comparative study of machine learning techniques in Healthcare. *2019 6th International Conference on Computing for Sustainable Global Development (INDIACom)*, pp. 455–460, New Delhi, India (13–15 March 2019). New York: IEEE.

2 Ganesan, M. and Sivakumar, N. (2019). IoT based heart disease prediction and diagnosis model for healthcare using machine learning models. *2019 IEEE International Conference on System, Computation, Automation and Networking (ICSCAN)*, pp. 1–5, Pondicherry, India (29–30 March 2019). New York: IEEE.

3 Pahwa, K. and Chauhan, S. (2021). Big data and machine learning in healthcare: tools & challenges. *2021 3rd International Conference on Advances in Computing, Communication Control and Networking (ICAC3N)*, pp. 326–330, Greater Noida, India (17–18 December 2021). New York: IEEE.

4 Ritter, F., Boskamp, T., Homeyer, A. et al., "Medical image analysis," *IEEE Pulse*, vol. 2, no. 6, pp. 60–70, 2011, https://doi.org/10.1109/MPUL.2011.942929.

5 Raj, R.J.S., Shobana, S.J., Pustokhina, I.V., Pustokhin, D.A., Gupta, D. and Shankar, K. (2020). "Optimal feature selection-based medical image classification using deep learning model in internet of medical things," in *IEEE Access*, vol. 8, pp. 58006–58017, 2020, https://doi.org/10.1109/ACCESS.2020.2981337.

6 Wenqian Chen, Sung Joon Park, Thomas Nok Hin Cheng, Nicholas Wai Hoi Lau, Liang Fa Khaw, Dan Lee-Lane, Xizhong Chen, Muhsincan Sesen (2022). "Pharmaceutical industry," in Sharifzadeh, M. *Industry 4.0 Vision for the Supply of Energy and Materials: Enabling Technologies and Emerging Applications*, Wiley, 2022, pp. 313–337, https://doi.org/10.1002/9781119695868.ch12.

7 Seyedan, M. and Mafakheri, F. (2020). Predictive big data analytics for supply chain demand forecasting: methods, applications, and research opportunities. *Journal of Big Data* 7: 53. https://doi.org/10.1186/s40537-020-00329-2.

8 Dahiwade, D., Patle, G. and Meshram, E. (2019). Designing disease prediction model using machine learning approach. *2019 3rd International Conference on Computing Methodologies and Communication (ICCMC)*, pp. 1211–1215, Erode, India. New York: IEEE.

9 Vayadande, K., Golawar, R., Khairnar, S. et al. (2022). Heart disease prediction using machine learning and deep learning algorithms. *2022 International Conference on Computational Intelligence and Sustainable Engineering Solutions (CISES)*, pp. 393–401, Greater Noida, India (20–21 May 2022). New York: IEEE.

10 Li, B., Li, W. and Zhao, D. (2013). Multi-scale feature based medical image classification. *Proceedings of 2013 3rd International Conference on Computer Science and Network Technology*, pp. 1182–1186, Dalian, China (12–13 October 2013). New York: IEEE.

11 Landset, S., Khoshgoftaar, T.M., Richter, A.N., and Hasanin, T. (2015). A survey of open source tools for machine learning with big data in the Hadoop ecosystem. *Journal of Big Data* 2: 24. https://doi.org/10.1186/s40537-015-0032-1.

12 Pynam, V., Spanadna, R.R., and Srikanth, K. (2018). An extensive study of data analysis tools (Rapid Miner, Weka, R Tool, Knime, Orange). *SSRG International Journal of Computer Science and Engineering (SSRG – IJCSE)* 5 (9): 4–11.

13 Nithya, B. and Ilango, V. (2017). "Predictive analytics in health care using machine learning tools and techniques," 2017 *International Conference on Intelligent Computing and Control Systems* (ICICCS), Madurai, India, pp. 492–499, https://doi.org/10.1109/ICCONS.2017.8250771.

5

A Review of Automated Sleep Stage Scoring Using Machine Learning Techniques Based on Physiological Signals

Santosh Kumar Satapathy[1], Poojan Agrawal[1], Namra Shah[1], Ranjit Panigrahi[2], Bidita Khandelwal[3], Paolo Barsocchi[4], and Akash Kumar Bhoi[4,5]

[1] *Information and Communication Technology, Pandit Deendayal Energy University, Gandhinagar, Gujarat, India*
[2] *Department of Computer Applications, Sikkim Manipal Institute of Technology, Sikkim Manipal University, India*
[3] *Department of General Medicine, Sikkim Manipal Institute of Medical Sciences, Sikkim Manipal University, Gangtok, Sikkim, India*
[4] *Wireless Networks (WN), Institute of Information Science and Technologies, National Research Council, Pisa, Italy*
[5] *Directorate of Research, Sikkim Manipal University, Gangtok, Sikkim, India*

5.1 Introduction

Historically, the general perception of sleep is a steady state of unconsciousness. Still, the actual truth is that it has an active and dynamic life cycle during which the brain is entirely functional [1]. Indeed, a proper sleep routine can significantly improve cardiac function and maintain appropriate mental restoration, attention, consolidation, and behavior. Sleep is an essential human physiological activity. Human beings spend approximately one-third of their lifetimes sleeping. Sleeping is necessary for the body to preserve its wellness and health. Sleep deprivation causes a reduction in agility and output during the daytime, reduces concentration, and adversely affects regular circadian rhythm [2, 3]. It also leads to mental and physical impairments, affects immunity, reduces cognitive functioning, and causes other health issues. Sleep is a continuous and dynamic process with a cyclic structural framework with a time course [4].

The rapid and nonrapid eye movements are considered to be the most potential stages of sleep. These stages of sleep are distinguished by unique neural activity traits [5, 6]. REM sleep underscores heightened neural engagement, while NREM sleep triggers reductions in metabolic rate, blood pressure, heart rate, and

Intelligent Techniques for Predictive Data Analytics, First Edition.
Edited by Neha Singh, Shilpi Birla, Mohd Dilshad Ansari, and Neeraj Kumar Shukla.

Published 2024 by John Wiley & Sons, Inc.

sympathetic activity, all the while accentuating parasympathetic activity [7]. The classification of sleep stages leans on well-defined criteria endorsed by authoritative bodies. Polysomnology (PSG) is emerging as a precise standard for dissecting sleep phases. PSG involves the simultaneous measurement of a variety of parameters, including electroencephalogram (EEG), electrocardiogram (ECG), EOG, EMG, blood oxygen quality, ventilation, and respiratory effort [8]. Typically enacted within a sleep laboratory, PSG data undergoes division into 30-second epochs, each precisely categorized into one of three NREM sleep stages (N1, N2, N3), REM sleep, or a wakeful state. The American Academy of Sleep Medicine (AASM) manual [9, 10] meticulously outlines the defining characteristics of each of these five sleep stages as follows:

Stage W: distinguished by regular or increased chin muscle activity, reading eye motions, or eye blinking [11].

Stage N1: distinguished by slowly moving eyes and a high probability of awakenings or arousals. Alpha bands must not surpass half of the broad spectral band, and the amplitude of the EEG signal should be in the frequency range of 2–7 Hz and remain less than 200 mV. N1 usually accounts for 5% of total sleep time [11].

Stage N2: Amrousals and awakening stage are less seen in N2 than in N1, and slowly moving eyes begin fading. N2 is awarded if there are one or more K-complexes or spindle trains in the last 50% of the previous or early 50% of the current epoch. N2 typically accounts for half of the sleep cycle in a single night [12].

Stage N3: It accounts for predominant slow waves and delta in the EEG signal and periodic arousals or awakenings. K-complex and Spindles may arise, and N3 is assigned if slow waves account for more than 20% of the period. N3 usually accounts for 20% of overall sleep. It depicts a deep sleep mode [13].

Stage REM: The dreaming stage is distinguished by the movement of the fast eye and active brain waves of N3. In REM, arousals and awakenings are common. The EEG exhibits mixed frequency and low voltage. REM is scored as such until a transition to stage W or N3. REM sleep accounts for one-fourth of total sleep time. Rechtschaffen and Kales (R&K) and the American Academy of Sleep Medicine (AASM) are two sets of guidelines for scoring sleep stages based on polysomnography (PSG) data [13].

The inadequate understanding of the underlying physiological mechanisms that shape the physiological signs acquired during sleep is one of the hurdles in sleep stage analysis (Table 5.1). Sleep patterns are known to change with age, but the source of these changes is unknown [14, 15]. The absence of apparent causality between physiological processes and observable signs complicates data interpretation. Furthermore, new and refined correlations are being uncovered in this dynamic area of research [16]. While human experts can still make diagnoses based on medical data, computational tools can help by detecting minor

Table 5.1 Sleep phases according to respective standards alongside descriptions.

Stages	R & K	AASM	Spectral band
WAKE (W)	W	W	
NREM	S1	N1	Alpha band
	S2	N2	Theta band
	S3	N3	Delta band
	S4		
REM	R	R	

discrepancies, speeding up analysis, and lowering expenses. These models can deliver a range of outputs, from event labeling to feature extraction and even diagnosis suggestions [17].

This research highlights the significance of diverse physiological signals housing insights into sleep stages and their classification [18]. These hold promise for diagnostic purposes, treatment monitoring, and the assessment of drug efficacy. However, realizing these advantages necessitates signal measurement and effective information extraction. The multitude of digital signal processing systems available has given rise to a lack of standardized methods for information retrieval, warranting the clarification of signals that carry sufficient diagnostic information [19]. To tackle this challenge, the study delves into the analysis of information extraction techniques across a spectrum of physiological signals, thereby shedding light on the inherent data information [20].

The entire review's findings regarding sleep scoring systems are rooted in these signals. Notably, each physiological marker corresponds to a single sleep stage, leading to the search for multiple signals for potential redundant but possibly uncorrelated information. As a result, much research has been conducted to investigate the automatic assessment of sleep stages through different combinations of signals [21].

5.2 Review of Related Works

5.2.1 Electroencephalogram (EEG)

The EEG captures the electrical activity of the brain, and the patterns of activity during different stages of sleep are diverse enough that they have been utilized to establish numerous sleep stage classification systems. Time-domain (TD) features [22], spectral features (Wave bands) [23], time-frequency (TF) [24] features, and nonlinear [25] features are all ways to extract sleep-related information from EEG data. Several classification approaches, including ensemble classification methods like Random Forest and Bootstrap Aggregating, K-means, and SVM, have been employed in sleep classification research to aid medical practitioners [26, 27]. A brief analysis of EEG signals for automatic sleep stage classification is discussed [28] (Figure 5.1; Table 5.2).

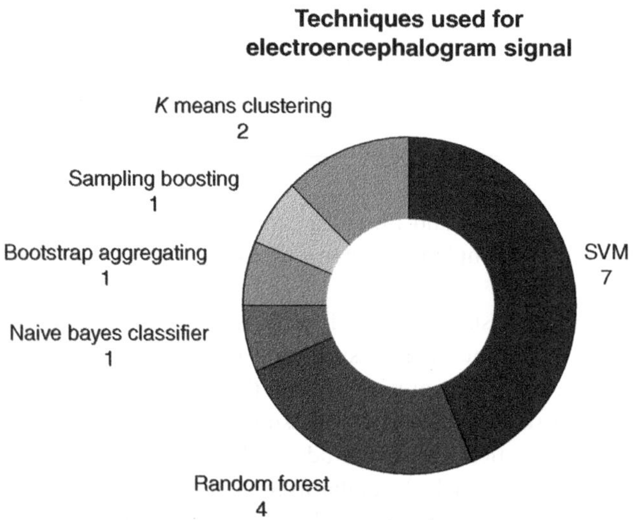

Figure 5.1 Employment of ML algorithms and alternative techniques in analyzing EEG signals.

Table 5.2 Overview of review outcomes for chosen studies employing EEG signals to aid sleep stage classification.

Existing studies	Dataset	Feature extraction and selection method	Classification method	Accuracy
[1]	Subjects: 20 Domain: healthy Age Group: 25 to 35 years	Linear evaluation technique	Support vector machine	Accuracy: Approximately 88% accuracy
[3]	EDF dataset (The benchmark Sleep)	Signal-sample-entropy and time-frequency wavelet filter	Support vector machine for a two-class problem	Accuracy: Approximately up to 98%
[2]	EDF dataset	Linear feature (Local Extrema)	Multiclass support vector machine	Accuracy: Approximately up to 98%

Table 5.2 (Continued)

Existing studies	Dataset	Feature extraction and selection method	Classification method	Accuracy
[4]	Subjects: 23 Domain: healthy adults Group: Male Age range: 23–45	Synchronization likelihood and relative entropy	Support vector machine and K-nearest neighbor technique	Accuracy: Approximately 93%
[6]	Sleep-EDF database	Nested five classes of sleep stage classification	Random forest (RFC)	Accuracy: Approximately 95%
[7]	Sleep-EDF database	Coupling techniques based on frequency features	Naive Bayes classifier	Accuracy: Approximately 94%
[8]	Sleep-EDF database	Wavelet transform techniques	Six-class classification based on Bootstrap aggregating (Bagging) from the Sleep-EDF database	Accuracy: Approximately 92%
[9]	Sleep-EDF database	Discrete Fourier Transform aka DFT	Random Forest aka RFC	Accuracy: Approximately 90%
[10]	Sleep-EDF dataset	Ensemble empirical mode decomposition	RUS Boost	Accuracy: Approximately 98%
[11]		Wavelet transform technique	Random forest aka RFC (six classes of sleep stage classification)	Accuracy: Approximately 91%
[12]	Sleep-EDF database	Time and frequency domain features and representation by entropy structural graph	K-means clustering technique	Accuracy: Approximately 96%
[13]	Subjects: 25 adults Domain: Sleep Apnea Dataset: provided by provided by St. VUHUCD	Two-stage multi-view learning approach	K-means clustering, SVM algorithm	Accuracy: Approximately 81%

(Continued)

Table 5.2 (Continued)

Existing studies	Dataset	Feature extraction and selection method	Classification method	Accuracy
[15]	Subjects: 25 individuals Age range: 40–60 years	Hybrid approach	RFC	Accuracy: Approximately 98%
[14]	Sleep-EDF database	Time-frequency domain features	Multiclass support vector machine	Accuracy: Approximately 88%
[16]	Subjects: 20 babies Domain: Region Croatia Age group: three months	Generalized zero-crossing methods	Support vector machine	Accuracy: 90%
[17]	Subjects: 28 Domain: Suspected Sleep Apnea Age range: 35–55	Recursive feature elimination technique	Binary support vector machine	Accuracy: Approximately 85%

5.2.2 Electrocardiogram (ECG)

ECGs without cardiac arrhythmias capture the electrical behavior of the heart, where their components can be identified individually by visual inspection. These ECG signals exhibit subtle changes consistent with autonomic nervous system (ANS) activities during sleep in particular times, e.g. prove that the decrease in volume occurs Consequently, PSG signals such as heart rate and respiratory rate change depending on the stage of sleep. These overlaps highlight the complex relationship between sleep status and physiological symptoms (Figure 5.2; Table 5.3).

5.2.3 Respiratory Effort and Electrooculography

Electrooculography (EOG) detects the REM stage by measuring the corneal retinal electrical potential, which can then be utilized to follow eye movements. The AASM recommends placing the EOG electrodes 1 cm from the left and right outside eye corners (canthus). This placement is straightforward, and patients can do it themselves, making it an effective tool for continuous sleep stage evaluation and long-term monitoring (Figure 5.3; Table 5.4).

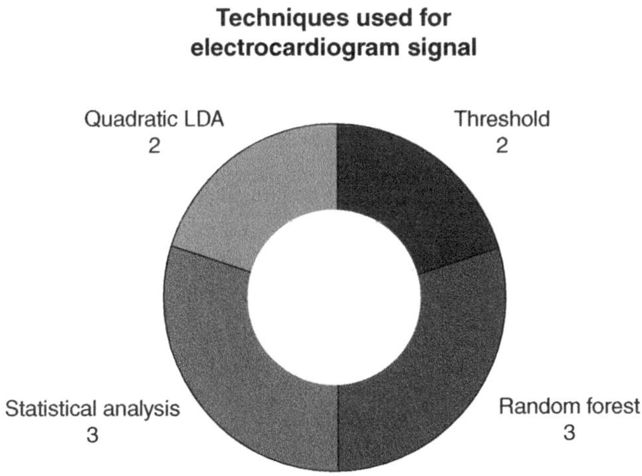

Figure 5.2 Utilization of ML algorithms and alternative techniques for analyzing ECG signals.

Table 5.3 Synopsis of review findings for handpicked studies employing ECG signals to facilitate sleep stage classification.

Wordsmith	Statistics and figures	Feature extraction and selection method	Classification method	Inferences
[18]	Dataset: CinC used for performance evaluation of the algorithm	HR statistics from ECG-derived respiration	Random Forest aka RFC	Accuracy: Approximately 87%
[19]	SLNEU database and PhyisioNet dataset	Morphological methods and techniques	Random Forest aka RFC	Accuracy: Approximately Up to 87%
[20]	Subjects: 51 (M: 37; F: 14) with obstructive sleep apnea (OSA) Surveyed at SNUH	HR parameters and statistical techniques, variability analysis	Threshold variability method	Accuracy: Approximately 88%
[21]	Subjects: 75 patients Domain: Sleep apnea Surveyed at SPTCMH	HR parameters and statistical techniques, spectral power	Statistical features	————

(*Continued*)

Table 5.3 (Continued)

Wordsmith	Statistics and figures	Feature extraction and selection method	Classification method	Inferences
[22]	EDF sleep	HR parameters and linear statistical techniques, NL frequency domain	Random Forest aka RFC	Accuracy: Approximately 89%
[23]	Database: SIESTA Domain: Apnea-ECG	HR parameters, spectral power	Threshold variability method	Accuracy: Approximately 58%
[24]	Subjects: 71 Domain: healthy women Group: postmenopausal	HR parameters, and linear statistical techniques, NL frequency domain	Statistical analysis	Not available
[25]	Subjects: 31 Group: Male	HR parameters and linear statistical techniques from ECG-derived respiration	Quadratic linear discriminant analysis	Accuracy: Approximately up to 76%
[26]	Subjects: 37	HR parameters and linear statistical techniques	Linear discriminant analysis	Not available
[27]	Subjects: 14 Domain: Healthy	HR parameters, and linear statistical techniques, and BP parameter	Statistical analytical techniques and comparison methods	Not available

5.2.4 Multiple Signals Integration (PSG)

Using multiple physiological signals provides redundant information since essential details in one signal may be missed but discovered in another [32]. As a result, PSG, which records EEG, ECG, EOG, and EMG signals, has become the threshold standard for detecting sleep problems [33]. PSG recordings are typically produced throughout the night. A sleep hypnogram is created by manually scoring sleep stages every 30 seconds by professional sleep experts. Physiological signals combine ECG, EEG, EOG, and EMG to build automatic sleep stage scoring systems [34] (Figures 5.4 and 5.5; Table 5.5).

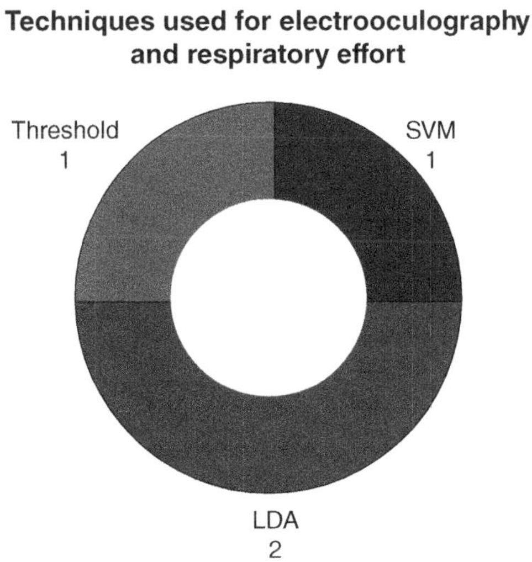

Figure 5.3 Application of ML algorithms and alternative techniques for EOG and respiratory effort analysis.

Table 5.4 Overview of sleep staging review outcomes for chosen studies utilizing EOG signals.

Wordsmith	Statistics and figures	Feature extraction and selection method	Classification method	Inferences
[28]	EOG Physio net DB	DW transform (DWT) coefficients and statistical analysis	Six classes problem approached with support vector machine	Accuracy: Approximately up to 92%
[29]	EOG	Spectral analysis	Five classes of sleep stage classification	Accuracy (Precision): Approximately 83%
[30]	Subjects: 48 Domain: Healthy (Respiratory effort)	Statistical analysis and spectral power	Quadratic linear discriminant analysis	Accuracy: Approximately 80%
[31]	Subjects: 265 Domain: Healthy (EOG)	Spectral analysis	Thresholds for percentage classes of sleep stage classification	Accuracy: Approximately 73%

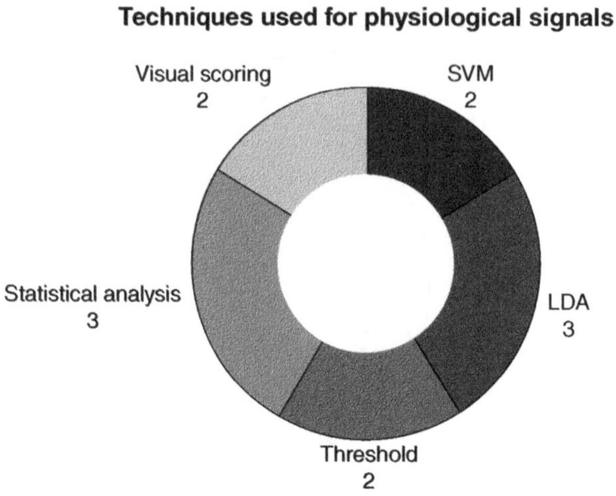

Figure 5.4 Deployment of ML algorithms and alternative techniques for processing PSG signals.

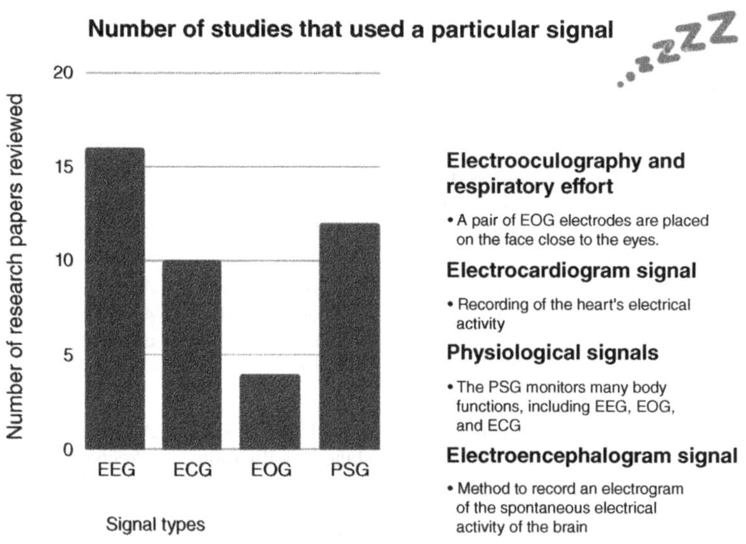

Figure 5.5 Graphical form of 42 research paper that used EOG, PSG, ECG, EEG signal.

5.3 Methodology

Preprocessing, feature selection and extraction, and classification models and algorithms are the primary steps in the quantitative sleep stage scoring procedure [42]. After feature extraction, some studies incorporate a

Table 5.5 Recapitulation of sleep staging review outcomes for chosen research incorporating multiple physiological signals.

Wordsmith	Statistics and figures	Feature extraction and selection method	Classification method	Inferences
[32]	Males: 664, i.e. (~66.67%), Females: 324, i.e. (~33.33%)	TP rate and TN rate	SVM	Accuracy (Precision): 82%
[33]	Newborns: 74 and Adults: 16: Total participants: 90	Spectral power on EEG-based signal, HR absolute HF, i.e., high-frequency component	Statistical analysis	Not available
[34]	Participants: 48	Spectral power is based on ECG signal, VM, and network analysis. Polysomnography: network analysis and time-frequency analysis	Linear discriminant analysis	Accuracy: Upto 80%
[35]	Database: sleep SIESTA	Time frequency and network analysis	Linear discriminant analysis	Accuracy: Approximately 80%
[36]	Subjects: 36 Domain: Healthy	Movement statistics, breathing rate statistics aka BR	Support vector machine	Accuracy: 81%
(Xi [30])	Participants: 115 adults; Respiratory effort, Domain: healthy	Statistical analysis	Linear discriminant analysis	Accuracy precision: Approximately 96 %
[37]	Healthy Subjects	Five-class sleep stage classification of sleep stages based on AASM rules	Statistical analysis	Not available
[38]	Participants: 230 Domain: Habitual Snorers And other 170 female participants (EEG and ECG)	Statistics and spectral power and visual scoring for PSG signals	Statistical analysis	Not available

(Continued)

Table 5.5 (Continued)

Wordsmith	Statistics and figures	Feature extraction and selection method	Classification method	Inferences
[39]	Participants: 542 children Domain: healthy participants from three to nine years being approximately	Five-class sleep stage classification of sleep stages based on AASM rules (visual scoring)	Visual scoring	Not available
[40]	Participants: 11 Infants; Domain: Healthy	EMG based signals	Threshold variability method	Not available
[41]	Participants: One participant having Parkinson's disease	Five-stage classification based on AASM rules (visual scoring)	Visual scoring technique	Not available

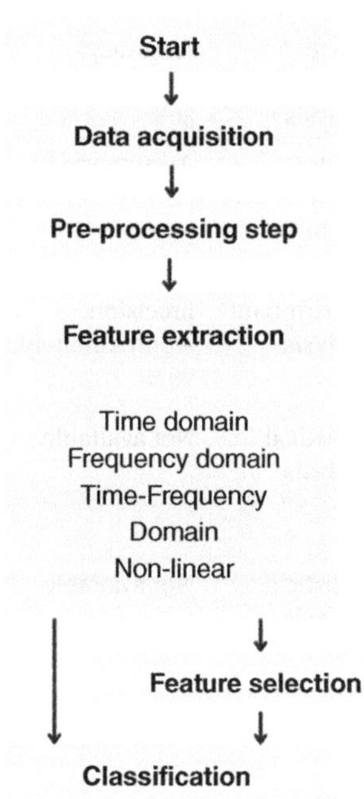

Figure 5.6 Process of automated sleep stage classification.

feature selection stage to select the most important and relevant features. The graphic below depicts the process of automatic sleep stage scoring (Figure 5.6) [43].

5.3.1 Preprocessing

Preprocessing involves cleaning raw EEG signals to remove artifacts and enhance the informative components. This is necessary to prevent incorrect or distorted results from various factors such as measurement noise, physiological variables, and pathological conditions. Despite using hardware filters while recording, outside sources can still contaminate EEG signals [44, 45].

5.3.2 Feature Extraction

Feature extraction encompasses the identification of informative attributes within a signal, which is essential for effective classification.

These derived features must possess independence and encapsulate the most distinctive data to avoid redundancy. To distill sleep stage insights from EEG recordings, a variety of signal processing techniques come into play. Extracted features are represented as in the form of either linear or nonlinear [44].

Time domain features offer simplicity and real-time adaptability as they elucidate the morphological characteristics of a signal. Frequency domain features on the contrary capture changes within EEG signals by transitioning time-based series into the frequency domain. Spectral estimation methods include parametric and nonparametric methodologies, such as periodogram and Welch, enabling the dissection of signal frequency components (Figure 5.7).

Within the nonlinear dynamics of EEG signals, nonlinear measures play a pivotal role in signal processing strategies. Notably, they find common application in feature extraction for sleep stage scoring. Entropy-based methods quantitatively gauge the irregularity and impurity of a time-domain signal [46]. In essence, lower entropy values signify more regular patterns within a signal window, while higher values indicate increased irregularity. Fractal-based methods generate a noise-like form by iteratively applying an operator on a geometric fragment. Despite their underlying rule-based behavior, certain instances exhibit noisy tendencies, adding complexity to their dynamics.

5.3.3 Feature Selection

After the feature extraction stage, some sleep stage classification methods, particularly multichannel ones, use feature selection algorithms to obtain a discriminative subset of characteristics. The goal is to reduce the number of features without sacrificing accuracy while avoiding overfitting and shortening computing time. Feature selection is handy when massive nighttime PSG records are used, and several types of characteristics are retrieved. Statistical strategies for determining the appropriate and discriminative subset of features have been developed in the literature. Sequential backward (SB) and sequential forward (SF) sampling are simple ways to examine features and determine the optimal subset. On the contrary, SF and SB sampling have limits and no backtracking. In contrast, metaheuristic algorithms such as genetic algorithm and particle swarm optimization are regularly used for feature selection [47, 48].

5.3.4 ML and DL for Sleep Disorder Detection

Recognizing the significance of sleep disorders, researchers have turned to innovative approaches, particularly ML and DL algorithms, to enhance detection methods. These computational techniques, which excel in pattern recognition

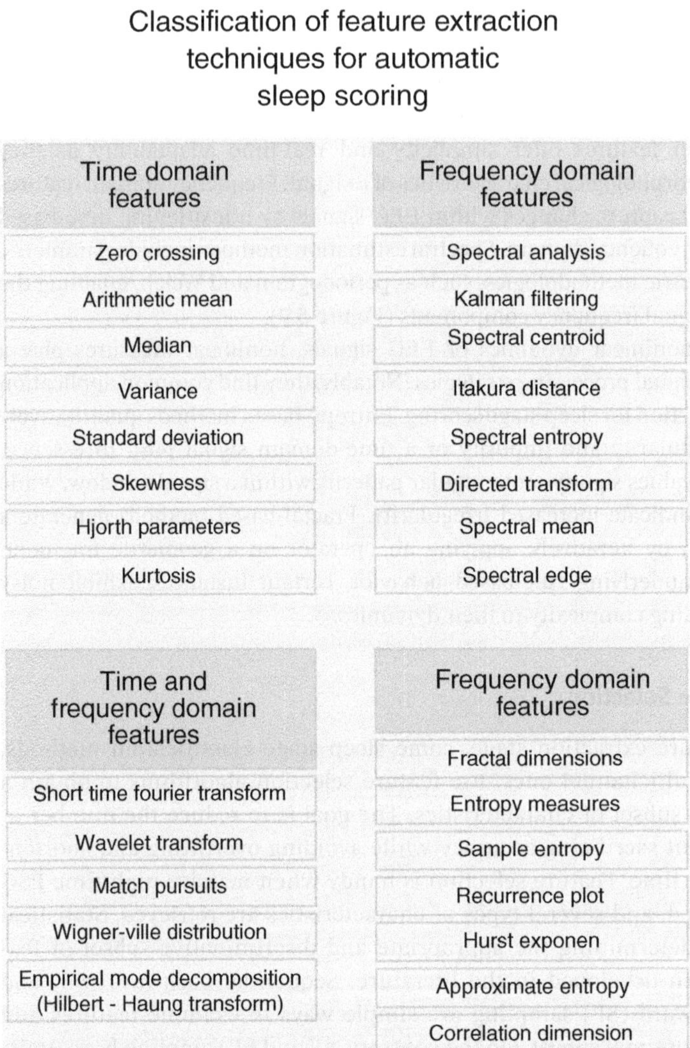

Figure 5.7 Categorization of feature extraction techniques for automated sleep scoring.

and data analysis, offer a promising means of identifying and diagnosing sleep disorders more accurately and efficiently. By analyzing extensive datasets encompassing sleep patterns, physiological signals, and other relevant parameters, ML and DL algorithms can reveal intricate patterns and correlations that might elude traditional diagnostic methods [49, 50].

5.3.4.1 Limitations of Traditional Methods

The limitations of traditional approaches for scoring sleep include the following:

Intrusiveness: Traditional methods often require patients to sleep in unfamiliar environments, affecting their natural sleep patterns and potentially leading to a phenomenon known as the "first-night effect."

Subjectivity: Manual scoring of sleep stages and events in PSG is prone to inter-scorer variability, leading to inconsistent results.

Accessibility: Sleep labs are limited in number and can be geographically distant, making it challenging for some patients to access these facilities.

Cost: PSG and related tests are expensive, involving not just equipment but also professional expertise for scoring and interpretation.

Limited Data: Traditional methods provide data from a single night in a controlled setting, potentially missing variations and fluctuations in sleep patterns over time.

5.3.4.2 Need for Automated Approaches (ML/DL)

The need for ML-/DL-based automated approaches for sleep pattern classification is essential for the following points (Figure 5.8).

Consistency and Value Chain: Machine learning (ML) and deep learning (DL) models provide methods that are practical and standardized, reducing variability among human scorers. This ensures an objective research design and assures consistent results in studies.

Noninvasive Sleep Monitoring: With sensor-enabled wearable devices, individuals' sleep patterns can be monitored in the comfort of their home. This preserves the natural sleep behavior of the subjects, allowing them to accurately replicate their typical sleep patterns.

Efficient Large-scale Studies: The automation provided by the ML and DL approaches facilitate the thorough screening of large numbers of people for sleep disorders. This early detection capability enables timely intervention and significantly reduces stress in specialty day hospitals, ultimately improving overall access to health care.

Deep-feature Extraction: ML and DL algorithms excel in extracting complex patterns that exist simultaneously across different physiological signals. This enhanced capability has the potential to reveal insights that traditional analytical methods may overlook.

Real-time Analysis and Feedback: Some automated systems provide users with real-time feedback, empowering them to make instant adjustments to improve their sleep This instant feedback this contributes to sustainable sexual improvement. Economic solutions once automated systems are developed,

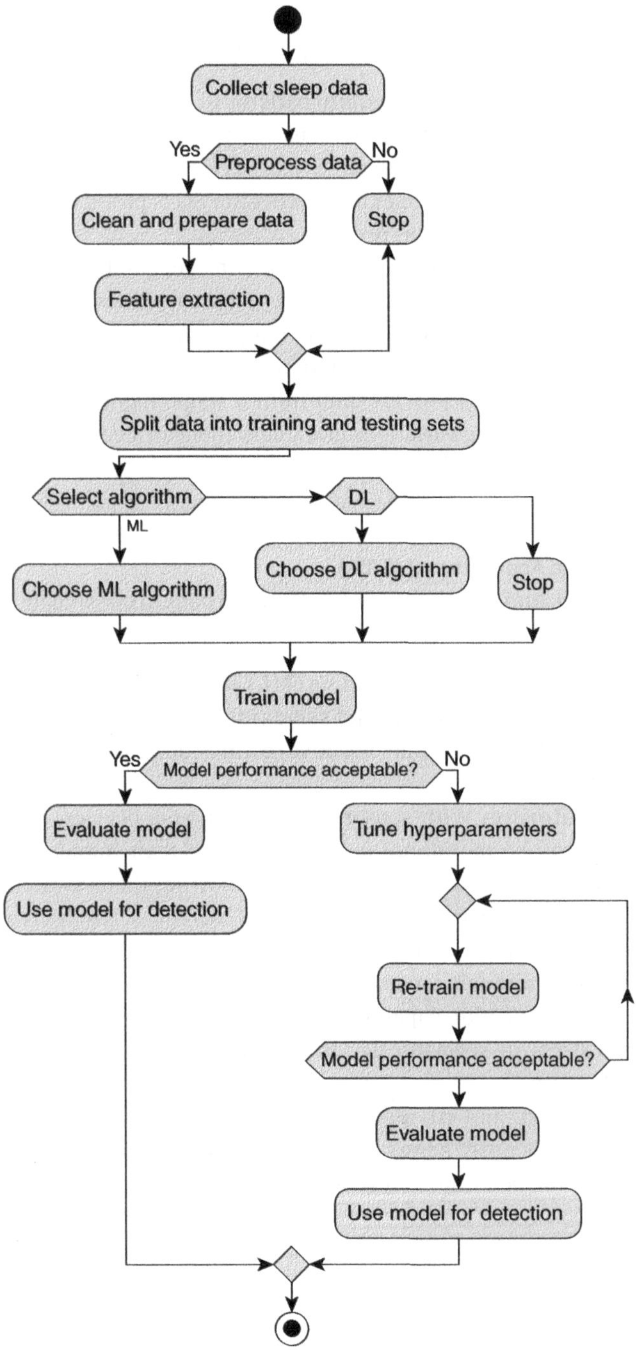

Figure 5.8 Comprehensive flowchart illustrating automatic sleep scoring through ML & DL approaches.

they offer a cost-effective alternative to hospitals with resource-intensive specialty days and repetitive testing. These potential cost savings hold significant promise for enhancing healthcare accessibility and affordability, with systems becoming notably more accessible and affordable.

Economical Solution: Once developed, automated systems offer a cost-effective alternative to resource-intensive specialty days hospitals and repetitive tests. These cost reductions can dramatically increase the affordability and availability of health care.

5.4 Conclusion

In this report, we delved into the transformative potential of ML and DL algorithms in revolutionizing sleep disorder detection. Through a comprehensive literature survey, we witnessed the remarkable impact of these algorithms on healthcare. The application of ML and DL in sleep disorder detection showcases their ability to process physiological signals from wearable devices, offering insights into disorders such as insomnia, sleep apnea, and narcolepsy. These algorithms' proficiency in deciphering intricate patterns within EEG, ECG, and other signals indicates their prowess in understanding sleep intricacies. Key to their success is feature selection, which highlights the significance of preprocessing data. The fusion of feature selection with model development has resulted in innovative models such as Deep Sleep, SleepNet, and NarcoNet. The complex architectures of CNNs and RNNs enable these models to discern sleep stages and disorder occurrences. Transitioning from model development to performance evaluation, metrics such as accuracy, sensitivity, and specificity validate the reliability of these algorithms in real-world clinical contexts. These metrics affirm the algorithms' effectiveness in addressing the multifaceted landscape of sleep disorders.

5.5 Future Work

The proposed study has led us to many avenues to contribute to our research work in sleep staging using PSG signals. In the future, we hope that some more new exciting frameworks and technology will be found by further analysis of the complications during the diagnosis of different types of sleep-related disorders. The methodology obtained in this thesis could be extended in several ways, for example:

- The performance of the proposed ensemble learning model can be extended by considering an appropriate selective aggregation technique that incorporates identifying the most effective features.

- In this thesis, we used only recordings of healthy controlled subjects and a few sleep-disordered subjects. In the future, we will focus on the large-scale sleep recordings of subjects who were affected by multiple sleep pathologies.
- Applying different wavelet decomposition techniques and other time-frequency features for a better way of analyzing discrimination of the changes in sleep behavior over the individual sleep stages.

References

1 Jiang, X., Zhao, J., Du, B., Yuan, Z. (2021). Self-supervised Contrastive Learning for EEG-based Sleep Staging, Harvard. *2021 International Joint Conference on Neural Networks (IJCNN)*, Shenzhen, China (18–22 July 2021). New York: IEEE, pp. 1–8.

2 Seifpour, S., Niknazar, H., Mikaeili, M., and Nasrabadi, A.M. (2018). A new automatic sleep staging system based on statistical behavior of local extrema using single channel EEG signal. *Expert Systems with Applications* 104: 277–293. https://doi.org/10.1016/j.eswa.2018.03.020.

3 Sharma, M., Goyal, D., Achuth, P.V., and Acharya, U.R. (2018). An accurate sleep stages classification system using a new class of optimally time-frequency localized three-band wavelet filter bank. *Computers in Biology and Medicine* 98: 58–75. https://doi.org/10.1016/j.compbiomed.2018.04.025.

4 Chriskos, P., Frantzidis, C.A., Gkivogkli, P.T. et al. (2018). Achieving accurate automatic sleep staging on manually pre-processed EEG data through synchronization feature extraction and graph metrics. *Frontiers in Human Neuroscience* 12: https://doi.org/10.3389/fnhum.2018.00110.

5 Long, X., Foussier, J., Fonseca, P. et al. (2014). Analyzing respiratory effort amplitude for automated sleep stage classification. *Biomedical Signal Processing and Control* 14: 197–205. https://doi.org/10.1016/j.bspc.2014.08.001.

6 Memar, P. and Faradji, F. (2018). A novel multi-class EEG-based sleep stage classification system. *IEEE Transactions on Neural Systems and Rehabilitation Engineering* 26 (1): 84–95. https://doi.org/10.1109/TNSRE.2017.2776149.

7 Dimitriadis, S.I., Salis, C., and Linden, D. (2018). A novel, fast and efficient single-sensor automatic sleep-stage classification based on complementary cross-frequency coupling estimates. *Clinical Neurophysiology* 129 (4): 815–828. https://doi.org/10.1016/j.clinph.2017.12.039.

8 Hassan, A.R. and Subasi, A. (2017a). A decision support system for automated identification of sleep stages from single-channel EEG signals. *Knowledge-Based Systems* 128: 115–124. https://doi.org/10.1016/j.knosys.2017.05.005.

9 da Silveira, T.L.T., Kozakevicius, A.J., and Rodrigues, C.R. (2017). Single-channel EEG sleep stage classification based on a streamlined set of statistical features in wavelet domain. *Medical & Biological Engineering & Computing* 55 (2): 343–352. https://doi.org/10.1007/s11517-016-1519-4.

10 Hassan, A.R. and Bhuiyan, M.I.H. (2017b). Automated identification of sleep states from EEG signals by means of ensemble empirical mode decomposition and random under sampling boosting. *Computer Methods and Programs in Biomedicine* 140: 201–210. https://doi.org/10.1016/j.cmpb.2016.12.015.

11 Hassan, A.R. and Bhuiyan, M.I.H. (2016). A decision support system for automatic sleep staging from EEG signals using tunable Q-factor wavelet transform and spectral features. *Journal of Neuroscience Methods* 271: 107–118. https://doi.org/10.1016/j.jneumeth.2016.07.012.

12 Diykh, M., Li, Y., and Wen, P. (2016). EEG sleep stages classification based on time domain features and structural graph similarity. *IEEE Transactions on Neural Systems and Rehabilitation Engineering* 24 (11): 1159–1168. https://doi.org/10.1109/TNSRE.2016.2552539.

13 Shi, J., Liu, X., Li, Y. et al. (2015). Multi-channel EEG-based sleep stage classification with joint collaborative representation and multiple kernel learning. *Journal of Neuroscience Methods* 254: 94–101. https://doi.org/10.1016/j.jneumeth.2015.07.006.

14 Bajaj, V. and Pachori, R.B. (2013). Automatic classification of sleep stages based on the time-frequency image of EEG signals. *Computer Methods and Programs in Biomedicine* 112 (3): 320–328. https://doi.org/10.1016/j.cmpb.2013.07.006.

15 Şen, B., Peker, M., Çavuşoğlu, A., and Çelebi, F.V. (2014). A comparative study on classification of sleep stage based on EEG signals using feature selection and classification algorithms. *Journal of Medical Systems* 38 (3): 18. https://doi.org/10.1007/s10916-014-0018-0.

16 Čić, M., Šoda, J., and Bonković, M. (2013). Automatic classification of infant sleep based on instantaneous frequencies in a single-channel EEG signal. *Computers in Biology and Medicine* 43 (12): 2110–2117. https://doi.org/10.1016/j.compbiomed.2013.10.002.

17 Koley, B. and Dey, D. (2012). An ensemble system for automatic sleep stage classification using single channel EEG signal. *Computers in Biology and Medicine* 42 (12): 1186–1195. https://doi.org/10.1016/j.compbiomed.2012.09.012.

18 Pini, N., Ong, J.L., Yilmaz, G. et al. (2022). An automated heart rate-based algorithm for sleep stage classification: validation using conventional polysomnography and an innovative wearable electrocardiogram device. *Frontiers in Neuroscience* 16: 974192. https://doi.org/10.3389/fnins.2022.974192.

19 Yücelbaş, Ş., Yücelbaş, C., Tezel, G. et al. (2018). Automatic sleep staging based on SVD, VMD, HHT and morphological features of single-lead ECG signal. *Expert Systems with Applications* 102: 193–206. https://doi.org/10.1016/j.eswa.2018.02.034.

20 Yoon, H., Hwang, S.H., Choi, J.W. et al. (2017). REM sleep estimation based on autonomic dynamics using R–R intervals. *Physiological Measurement* 38 (4): 631–651. https://doi.org/10.1088/1361-6579/aa63c9.

21 Liu, S., Teng, J., Qi, X. et al. (2017). Comparison between heart rate variability and pulse rate variability during different sleep stages for sleep apnea patients. *Technology and Health Care* 25 (3): 435–445. https://doi.org/10.3233/THC-161283.

22 Xiao, M., Yan, H., Song, J. et al. (2013). Sleep stages classification based on heart rate variability and random forest. *Biomedical Signal Processing and Control* 8 (6): 624–633. https://doi.org/10.1016/j.bspc.2013.06.001.

23 Kesper, K., Canisius, S., Penzel, T. et al. (2012). ECG signal analysis for the assessment of sleep-disordered breathing and sleep pattern. *Medical & Biological Engineering & Computing* 50 (2): 135–144. https://doi.org/10.1007/s11517-011-0853-9.

24 Virtanen, I., Ekholm, E., Polo-Kantola, P., and Huikuri, H. (2007). Sleep stage dependent patterns of nonlinear heart rate dynamics in postmenopausal women. *Autonomic Neuroscience* 134 (1–2): 74–80. https://doi.org/10.1016/j.autneu.2007.01.010.

25 Redmond, S.J., de Chazal, P., O'Brien, C. et al. (2007). Sleep staging using cardiorespiratory signals. *Somnologie - Schlafforschung und Schlafmedizin* 11 (4): 245–256. https://doi.org/10.1007/s11818-007-0314-8.

26 Redmond, S.J. and Heneghan, C. (2006). Cardiorespiratory-based sleep staging in subjects with obstructive sleep apnea. *IEEE Transactions on Biomedical Engineering* 53 (3): 485–496. https://doi.org/10.1109/TBME.2005.869773.

27 Trinder, J., Kleiman, J., Carrington, M. et al. (2001). Autonomic activity during human sleep as a function of time and sleep stage. *Journal of Sleep Research* 10 (4): 253–264. https://doi.org/10.1046/j.1365-2869.2001.00263.x.

28 Rahman, M.M., Bhuiyan, M.I.H., and Hassan, A.R. (2018). Sleep stage classification using single-channel EOG. *Computers in Biology and Medicine* 102: 211–220. https://doi.org/10.1016/j.compbiomed.2018.08.022.

29 Liang, S.F., Kuo, C.E., Lee, Y.C. et al. (2015). Development of an EOG-based automatic sleep-monitoring eye mask. *IEEE Transactions on Instrumentation and Measurement* 64 (11): 2977–2985. https://doi.org/10.1109/TIM.2015.2433652.

30 Long, X., Fonseca, P., Foussier, J. et al. (2014). Sleep and wake classification with actigraphy and respiratory effort using dynamic warping. *IEEE Journal of Biomedical and Health Informatics* 18 (4): 1272–1284. https://doi.org/10.1109/JBHI.2013.2284610.

31 Virkkala, J., Hasan, J., Värri, A. et al. (2007). Automatic sleep stage classification using two-channel electro-oculography. *Journal of Neuroscience Methods* 166 (1): 109–115. https://doi.org/10.1016/j.jneumeth.2007.06.016.

32 Pini, N., Ong, J.L., Yilmaz, G. et al. (2021). An automated heart rate-based algorithm for sleep stage classification: validation using conventional PSG and innovative wearable ECG device. *medRxiv* 16 (974192): 1–18.

33 Takatani, T., Takahashi, Y., Yoshida, R. et al. (2018). Relationship between frequency spectrum of heart rate variability and autonomic nervous activities during sleep in newborns. *Brain and Development* 40 (3): 165–171. https://doi.org/10.1016/j.braindev.2017.09.003.

34 Fonseca, P., Long, X., Radha, M. et al. (2015). Sleep stage classification with ECG and respiratory effort. *Physiological Measurement* 36 (10): 2027–2040. https://doi.org/10.1088/0967-3334/36/10/2027.

35 Helland, V.C., Gapelyuk, A., Suhrbier, A. et al. (2010). Investigation of an automatic sleep stage classification by means of multiscorer hypnogram. *Methods of Information in Medicine* 49 (05): 467–472. https://doi.org/10.3414/ME09-02-0052.

36 Willemen, T., Van Deun, D., Verhaert, V. et al. (2014). An evaluation of cardiorespiratory and movement features with respect to sleep-stage classification. *IEEE Journal of Biomedical and Health Informatics* 18 (2): 661–669. https://doi.org/10.1109/JBHI.2013.2276083.

37 Kishi, A., Yasuda, H., Matsumoto, T. et al. (2011). NREM sleep stage transitions control ultradian REM sleep rhythm. *Sleep* 34 (10): 1423–1432. https://doi.org/10.5665/SLEEP.1292.

38 Kesek, M., Franklin, K.A., Sahlin, C., and Lindberg, E. (2009). Heart rate variability during sleep and sleep apnoea in a population based study of 387 women. *Clinical Physiology and Functional Imaging* 29 (4): 309–315. https://doi.org/10.1111/j.1475-097X.2009.00873.x.

39 Montgomery-Downs, H.E., O'Brien, L.M., Gulliver, T.E., and Gozal, D. (2006). Polysomnographic characteristics in normal preschool and early school-aged children. *Pediatrics* 117 (3): 741–753. https://doi.org/10.1542/peds.2005-1067.

40 Estévez, P.A., Held, C.M., Holzmann, C.A. et al. (2002). Polysomnographic pattern recognition for automated classification of sleep-waking states in infants. *Medical & Biological Engineering & Computing* 40 (1): 105–113. https://doi.org/10.1007/BF02347703.

41 Tracik, F. and Ebersbach, G. (2001). Sudden daytime sleep onset in Parkinson's disease: polysomnographic recordings. *Movement Disorders* 16 (3): 500–506. https://doi.org/10.1002/mds.1083.

42 Satapathy, S.K. and Loganathan, D. (2023). Automated classification of multi-class sleep stages classification using polysomnography signals: a nine- layer 1D-convolution neural network approach. *Multimedia Tools and Applications* 82 (6): 8049–8091. https://doi.org/10.1007/s11042-022-13195-2.

43 Satapathy, S.K., Bhoi, A.K., Loganathan, D. et al. (2021a). Machine learning with ensemble stacking model for automated sleep staging using dual-channel EEG signal. *Biomedical Signal Processing and Control* 69: 102898. https://doi.org/10.1016/j.bspc.2021.102898.

44 Satapathy, S.K. and Loganathan, D. (2021b). Prognosis of automated sleep staging based on two-layer ensemble learning stacking model using single-channel EEG signal. *Soft Computing* 25 (24): 15445–15462. https://doi.org/10.1007/s00500-021-06218-x.

45 Satapathy, S.K. and Loganathan, D. (2022). Automated classification of sleep stages using single-channel EEG. *International Journal of Information Retrieval Research* 12 (2): 1–19. https://doi.org/10.4018/IJIRR.299941.

46 Acharya, U.R., Sudarshan, V.K., Adeli, H. et al. (2015). Computer-aided diagnosis of depression using EEG signals. *European Neurology* 73 (5–6): 329–336. https://doi.org/10.1159/000381950.

47 Erguzel, T.T., Ozekes, S., Tan, O., and Gultekin, S. (2015). Feature selection and classification of electroencephalographic signals: an artificial neural network and genetic algorithm based approach. *Clinical EEG and Neuroscience* 46 (4): 321–326.

48 Surantha, N., Lesmana, T.F., and Isa, S.M. (2021). Sleep stage classification using extreme learning machine and particle swarm optimization for healthcare big data. *Journal of Big Data* 8 (1): 14. https://doi.org/10.1186/s40537-020-00406-6.

49 Korkalainen, H., Leppanen, T., Aakko, J. et al. (2019). Accurate deep learning-based sleep staging in a clinical population with suspected obstructive sleep apnea. *IEEE Journal of Biomedical and Health Informatics* 1–1. https://doi.org/10.1109/JBHI.2019.2951346.

50 Sekkal, R.N., Bereksi-Reguig, F., Ruiz-Fernandez, D. et al. (2022). Automatic sleep stage classification: from classical machine learning methods to deep learning. *Biomedical Signal Processing and Control* 77: 103751. https://doi.org/10.1016/j.bspc.2022.103751.

6

Predictive Analytics for Marketing and Sales of Products Using Smart Trolley with Automated Billing System in Shopping Malls Using LBPH and Faster R-CNN

Balla Adi Narayana Raju[1], *Deepika Ghai*[1], *Suman Lata Tripathi*[1], *Sunpreet Kaur Nanda*[2], *and Sardar M.N. Islam*[3]

[1] *School of Electronics and Electrical Engineering, Lovely Professional University, Phagwara, India*
[2] *Electronics and Telecommunication Engineering Department, P.R. Pote Patil College of Engineering and Management, Sant Gadge Baba Amravati University, Amravati, India*
[3] *Institute for Sustainable Industries and Liveable Cities, Victoria University, Melbourne, Australia*

6.1 Introduction

Over the past decade, the internet and the economy have flourished simultaneously, leading to the birth of an era of hassle-free shopping from the convenience of one's living room. Internet technologies have always had the edge in tracking customers' purchase habits and exploiting them. However, traditional retail stores have not had a similar level of targeted advertisement and curated shopping experience [1, 2]. IoT technology in retail stores has remained relatively untouched by automation, image processing, and artificial intelligence (AI) advancements. Considering that supermarkets are where customers get their constant supply of daily use items, such as food products, garments, electronics devices, and household appliances [3, 4]; therefore, any amount of advancement based on customer preferences can cause a direct spike in the sales and perception of the product. Supermarket faces constant challenges like scanning many barcode items [2, 5], which increases the time and results in queuing delay as shown in Figure 6.1. There is a need to reshape the billing process by implementing self-service check-outs. Recent advancements in chip technology have resulted in several realistic methods for new, hassle-free applications [6]. Barcode scanner takes a long time, but this can be avoided by attaching RFID tags to each item, which can then be scanned by RFID scanners that are both authentic and compact [4–7].

Intelligent Techniques for Predictive Data Analytics, First Edition.
Edited by Neha Singh, Shilpi Birla, Mohd Dilshad Ansari, and Neeraj Kumar Shukla.

Published 2024 by John Wiley & Sons, Inc.

Figure 6.1 Problems in supermarkets. *Source:* Yewatkar et al. [4]/Reproduced with permission from Elsevier.

6.2 Major Contributions

In this chapter, an efficient and effective trolley-based model for monitoring the stores as well as the customers' buying habits for a prediction of the items for better profitability is proposed. The proposed trolley is highlighted because it offers customer face recognition, automatic billing, customer tracking, and store classification. The smart trolley is built with the RFID reader, pi-camera, ultrasonic sensors, ATmega328P controller, and Broadcom BCM2837 Advanced Risk Machine7 (ARM7) quad-core processor as the main components [8, 9]. The customer in the store can scan each RFID tag on the item with an RFID reader and view information such as the product name and price along with the total balance on his/her mobile phone application. Customers have more flexibility since the bill is changed with each scanning of items and removed with scanned items. This trolley also helps customers to see their overall bill, which aids in budget control [10–12]. The management interface system (MIS) obtains information about the products and their costs, as well as alerts the store manager when new items need to be added to the racks on time. The face recognition used initially with the pi-camera will detect the customer details and send it to MIS for a better shopping experience and interests to get customers' buying patterns. Ultrasonic sensors HC-SR04 used on either of the sides of the trolley will help in deciding the path according to the customer's movements for a hassle-free experience [13–15]. Finally, people will have to pay their bills online instead of standing in long queues.

A unified trolley system is proposed in this chapter by combining IoT, image processing, ultrasonic sensing, and RFID technology. RFID has become a rapidly growing technology in economical industries for item-level scanning [5, 16, 17]. In the coming days, these RFID tags are likely to be the most widely used. The 32-bit ARM7 processor and ATmega328 controller are the core of the trolley, which interfaces with the MIS, mainframe, and mobile. These ARM7 and ATmega328 microcontrollers are the most used embedded system platforms due to their fast and efficient specifications. The proposed trolley aims to provide customers and store managers with a hassle-free experience. This technology also provides a flexible, easily scalable, robust, cost-effective, and technology-oriented shopping experience, thereby reducing the queuing time.

The major challenges faced by the proposed trolley system in the supermarket setting for automation, detection, recognition, and analysis are the exertion of heavy loads on the self-propelled system a heavy load-hauling usually takes up high torque and overhauling could damage the motors. Also, the load in each cart is variable (this can though be accomplished with pulse width modulation [PWM]). So, managing the speed of the trolley is a major concern for automatic movement in the system.

The detection of the customer is also a major challenge [18–22]. Detecting the correct client and getting the data from the database to provide a better shopping experience accurately is concerning as errors caused by not recognizing the face properly due to the facial features can cause wrong recommendations. However, by using the best face recognition algorithms, facial analysis can be performed without any mismatch [23, 24].

The backgrounds in the real-life scenarios are much more complex compared to the technical implementation datasets. Racks with items can be of uneven form thus being hard to detect and analyze if there are a large number of items placed. Training the convolutional neural network (CNN) with different store scenarios can reduce fault prediction [21–23].

Building up this kind of new framework and using it in the store will require a huge number of assets and information at the underlying stages to prepare the model. Heaps of testing stages need to be done to improve the procedure stream of the trolley from facial recognition, human following, and tag detection to monitoring the surroundings of the store [25–28]. Another challenge is making users take up these changes in a positive light, creating initial data that would help in future recommendations following principles of data security and privacy.

The rest of this chapter is organized as follows: Section 6.2 represents the related work in the field of trolley-based systems. Section 6.3 provides the implementation details of the proposed method. Section 6.4 discusses the results of various modules in the trolley. Finally, concluding remarks and promising research directions for the future are given in Section 6.5.

6.3 Related Work

Researchers have proposed numerous methods for trolley-based systems. In 2011, Agarwal et al. [6] built an RFID-based automatic shopping cart that allows the customer a superior approach to shopping in-store environment. This shopping trolley comprises a touch screen for the customer graphical interface attached to the cart, a client card, and an item reader for scanning the labels of the items placed in the cart, which makes it customer-intuitive.

In 2013, McBeath [7] formulated that retailers need to adapt to another technology to persevere the complexity of items day by day. They require exact, real time, and item-level pictures of stock. RFID has exhibited by improving store deals by 20–30%. JC Penney used RFID to increase stock accuracy from 75% to 99% in the gathering of products.

Kumar et al. [8] proposed an autonomous supermarket trolley system where RFID technology, ZigBee, and microcontrollers are used. This system has a start button for making the trolley active when an item with an RFID tag is placed in the trolley. The RFID reader reads the tag ID, and if the number matches the database, the product information tagged to that ID is displayed in the liquid crystal display (LCD). This display also has an end shopping button, when this button is pressed the total list of items present in the trolley is then sent to the master PC via ZigBee. There is also an option in the LCD to remove a particular item from the cart. This bill will be updated according to the customer's choice. The limitation of this system is that payments at the end are done on the billing counter for which the user should again waste sometime in the billing of items and the trolley is not smart enough to interact with the surroundings of the store.

In 2014, Kamble et al. [9] used RFID technology to build a multiprocessing shopping cart system. The item information is collected at the time of billing, and the RFID scanner is used to read it. This shopping cart system is made up of the following three parts: a display component and a user interface, a server communication component, and an automated billing component, all of which work together to create a smart trolley that significantly reduces billing time. Certain problems, such as stock maintenance, are not addressed in this scheme. That is if the substance that has been dropped in the trolley is not read and the required remedial action is not addressed.

Yewatkar et al. [4] discussed the components required for building a smart cart system with automatic billing, product information, and product recommendation using RFID, and ZigBee with anti-theft using AT89S52 microcontroller, electrically erasable programmable read-only memory (EEPROM), and LCD. The high data rate, voltages, communication protocol, and frequency requirements are shown, as well as the information on the inter-integrated circuit (I2C) protocol between the microcontroller and EEPROM. Memory size and the need for more sophisticated communication protocols are both disadvantages of this configuration.

In 2015, Bhatwal et al. [10] proposed an automated self-billing system using RFID technology which consists of a display panel, an RFID reader, and a master computer. The passive tags are initially scanned by the reader as shown in Figure 6.2, which is then shown on the touch display panel if the user wants to remove the item then he/she can remove the item using the cancellation button in the display panel. The display panel also prints the bill at the end which the customer should pay using cash or card. The drawback of this system is the display panel that the author has described is a master computer which in reality, the billing process is complex.

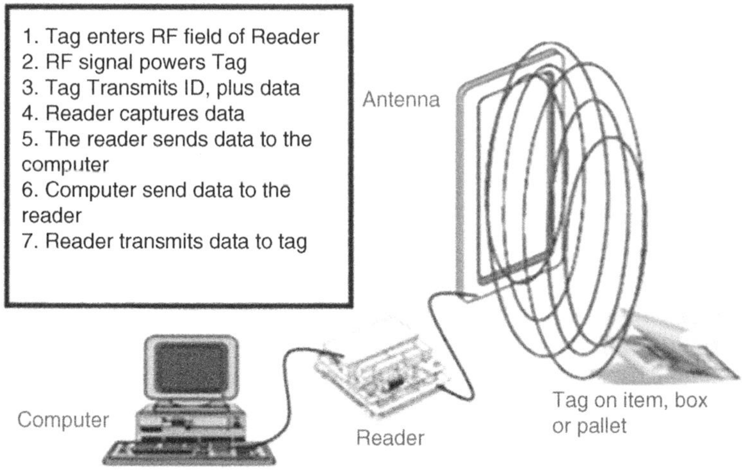

Figure 6.2 Working of RFID tag. *Source:* Bhatwal and Chamoli [10]/International Journal of Emerging Trends & Technology in Computer Science.

In 2016, Shraddha et al. [1] proposed an IoT-based interactive trolley for shopping malls that uses RFID tags and includes an RFID reader, microcontroller, LCD, and an Espressif (ESP) module in the cart. The ESP offers real-time details on what's going on in the stores from anywhere. The major benefit of this system is there is no need for initial training.

Lee et al. [11] projected a mobile robot that can track and lead humans by using wearable devices to detect user position and behavior, as shown in Figure 6.3. This chapter suggests the prototype of a small robot that can follow the user by using

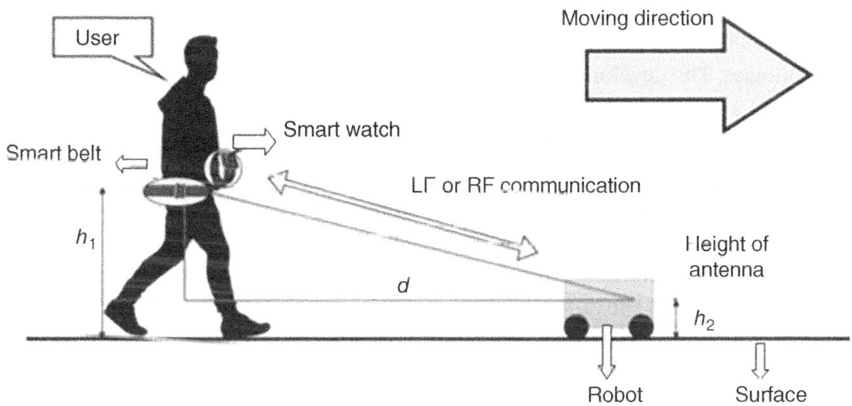

Figure 6.3 Smart mobile robot. *Source:* Lee et al. [11]/with permission of IEEE.

the smartwatch and belt to track the person using radio frequency (RF) communication with a low-frequency (LF) transmitter. The drawback of this idea was that tracking or finding a person was difficult and was not so accurate.

Chiang et al. [12] proposed a smart supermarket trolley that included a face recognition feature for store customers. The customers are also assisted with product recommendations based on their history of purchases. In this way, buying efficiency can be increased even further, and shopping time can be minimized. The algorithms for cart navigation and face recognition, on the contrary, are extremely complicated.

Wu et al. [13] suggested the concept of an intelligent shopping assistant system (ISAS) that could recognize and navigate the trolley consistently. Customers can see their shopping list and guide the shopping trolley to get items with this model's additional app functionality. Customers can see their shopping list and direct the shopping trolley to get products using the app which is also an additional feature of this app. Product suggestions and discount details are also included in the app. This model has the benefit of providing a hands-free shopping experience for the least amount of time spent at the shopping mall.

In 2018, Sarala et al. [29] designed a smart electronic trolley for shopping malls to reduce the time to purchase products. The barcode scanner on the electronic trolley scans the results to identify the items and connects to the shop's server over the Internet. In addition, it includes LCDs that inform customers of the number of items and their total cost and a barcode scanner that determines the result and updates the bill. A swiping machine is available so that customers can pay their bills using credit or debit cards.

Keerthana et al. [30] developed a system in which visually impaired people can enjoy their shopping experience and purchase the products they want. A specific personal gadget was designed for the identification of products and navigation. This project demonstrates how wireless personal gadgets are used in everyday life. It provides visually impaired people with a smart environment by using RFID and Zigbee technologies. The developed product enables visually impaired people to shop independently and not depend on others. You do not need any special training to use it, it is easy to use, and it is economical. The fully automated product overcomes the hesitation of purchasing their shopping needs and gives them confidence.

Devipriya et al. [31] stated that to enhance the quality of life for visually impaired and blind people, many methods have been proposed to date. However, purchasing products in a supermarket without others' assistance is still a big challenge for them. The purpose of this chapter is to propose a smart store assistant for visually impaired people. The assistant addresses three major concerns, namely, finding the product, identifying the product, and purchasing the product. The system has three modules: product identifier, smart glove, and smart trolley. RFID is used to identify products in the product identifier module. By scanning the product with the smart glove, the buyer receives audio instructions that explain the product specifications. As soon as the product is added to the cart, an automated billing system provides a final bill for the purchased products. An indoor navigation system will be developed in the future to locate the desired product from the user's location. This work demonstrates the design and implementation of a working prototype model. It is possible to improve or replace the low-range RFID reader (EM-18) in the present work with a high-range RFID reader if this prototype is implemented in real time.

In 2019, Gunawan et al. [32] developed an automatic moving trolley, to solve the inconvenience and time wasted by customers looking for the desired product in the store. The purpose of a shopping trolley or shopping cart when shopping in a supermarket is to transport items to the cashier and not to take them outside. This chapter describes the hardware and software designs of a smart trolley system based on a two-wheeled mobile robot developed previously. Various sensors, IOIO microcontrollers, and Android smartphones are used in this chapter. The Android smartphone controls the robot by sending a signal to the microcontroller paired with the actuator and monitors the situation with the camera of the smartphone. A Navisens indoor positioning system is also included in this system, which detects the user's location using the smartphone's accelerometer and gyroscope. The Navisens framework-based smart trolley system can move and display its location to the user as the result of testing on robot navigation. Navisens provides a better indoor positioning system than any place because it updates its estimated position using the dead reckoning system, accelerometers, and gyroscopes instead of Wi-Fi fingerprints. Despite being good at a distance of 1 and 3 meters, the smart trolley is not as stable at distances over 3 meters. It will be studied in the future whether the smart trolley system requires an automated calibrator to ensure the trolley does not depart from the path based on the analyses.

In 2020, Anitha et al. [33] stated that the objective of this chapter is to eliminate the problem of standing in line and wasting time while shopping. This study proposes a smart trolley billing system based on RFID tags that audits purchased products and makes payments online automatically. The cart identifies and scans products automatically, and the final billing is handled right there on the trolley. In addition, it provides customers with centralized and automated billing systems using RFID, making it a reasonable and profitable smart shopping cart. This article provides a time saving, commercially oriented, technology-based system that enhances customer shopping experiences. The proposed system also suggested products based on user purchase history. With the proposed system, every product in the store should have an RFID tag, and every cart should have an RFID reader, which will save time and simplify shopping. This chapter has shown some challenges with smart shopping, i.e. sometimes items cannot be detected due to their tag orientations, sizes, and shapes. These interactions are usually quite expensive. These drawbacks have been addressed with this application. An LCD screen displays the automated billing amount on the smart trolley.

Das et al. [34] have proposed a smart trolley that handles both shopping and billing. Shopping can be thrilling and alluring, but it can also be time-consuming and boring when you have to wait in long lines to pay your bill. With this system, the customer can walk into the shop right away, purchase products using the smart trolley, and walk out. He receives the e-bill by mail and can view the details of his purchase online. To manage product and customer information, it requires an Arduino board, RFID reader, RFID tag, LCD, ESP8266 Wi-Fi module, and a website that can be accessed anywhere in the world. This IOT-based system allows the trolley to interact with a global network.

In 2021, Shailesh et al. [35] proposed a smart trolley for shopping. Weekends and holidays usually see a huge rush in markets, especially in metropolises. Huge discounts and offers make the rush even greater. We get a trolley in shopping malls in which we place all the items intended to be purchased from different shelves. After putting all the items into the trolley, we reach the counter for billing, leading to long queues. To solve this problem, a barcode reader and LCD screen are attached to the trolley. When the item is placed in the trolley, the barcode automatically detects the product's barcode and displays the item's details on a screen. The cost of the item is then added to the bill and the entire work is done with the help of Raspberry Pi. This practice will eliminate long queues at the counters because the trolley will do all the billing.

In 2022, Shete et al. [36] explained that not being able to see well or at all is just the tip of the iceberg for visually impaired people. Those groups face a whole host of societal challenges, stigmas, fears, and beliefs that make them the minority, and we do not give enough attention to what they face. It is incorrect to ignore the larger picture when chasing buzzwords like artificial intelligence and robotics to create complex technologies to solve mundane problems, but it is also essential to remember that mankind can use the bigger picture to save the rest of the planet. From navigating a path to shopping at the mall, those with visual impairments face a variety of difficulties. With braille-assisted buttons and haptic feedback control, this chapter presents the idea of a smart shopping cart that helps visually impaired individuals navigate the various aisles of a shopping mall and purchase items without the need for human assistance.

Patil et al. [37] explained that there are several modern shopping malls (supermarkets) where we can purchase a wide range of goods, such as glass, toys, kitchen sets, groceries, and decorative items. Discounts, home delivery, and other factors make people shop. Due to busy schedules, boredom, and dissatisfaction with the services provided at the billing counter, customers have to wait in long lines, especially on weekends. For COVID-19, we have implemented an effective method of social distancing to avoid these problems. As a solution to this unique customer problem, especially when it comes to billing, the system proposed is the

Smart trolley, which scans items as they are put in, and displays the purchase amount and item in parallel on the LCD screen. It can scan a large number of items using RFID technology, saving both malls and consumers time.

Even though many researchers have proposed a large number of ideas [1, 4, 6–13, 29–37] a single unified approach needs to be done that fits all applications [38]. In this chapter, a generic trolley-based framework is proposed that uniformly describes the smart way of shopping in stores with a trolley interactive mechanism for showing the path and predicting the customer purchasing patterns.

6.4 Proposed Methodology

The trolley architecture comprises four modules as shown in Figure 6.4. Each module has its advantages from getting customer details, and product details and classifying the store surroundings to the customer following. There are also image classification techniques used to monitor the surroundings for product identification and customer purchase pattern prediction for marketing strategies.

6.4.1 Module 1 – Face Recognition

Customer's faces are identified by the trolley using the camera module. The face is coordinated with the dataset of the store to check whether the face is present in the database or not. If there are matching facial characters found, then the trolley will be activated and the data corresponding to the application ID of the customer will be retrieved into the client's application with a detailed chart of his/her interest and previous transactions. The client can also log in manually using his/her front-facing camera, receiving a quick response (QR) that can be scanned at the trolley. In this event, if the face is not found in the dataset, at that point the face is added to the database (DB), and another ID is generated and activated with the new customer ID. The identification of the facial features has been done using a highly accurate face recognition algorithm, i.e. local binary pattern histogram (LBPH) of the OpenCV library to avoid the error associated with the detection of faces [14]. Figure 6.5 shows the flowchart of the face recognition process.

The following are the stages involved in the face recognition system:

a) **Image Acquisition:** It is where input images are gathered.
b) **Face Detection:** The technology that identifies human faces in digital images is essentially computer technology that is used in a variety of applications.
c) **Face Recognition:** It is a biometric technique for identifying an individual by comparing advanced image data to the person's stored record.
d) **Image Pre-processing Techniques:** There is a need for preprocessing techniques to improve the quality of input images. The color-to-gray-scale image conversion

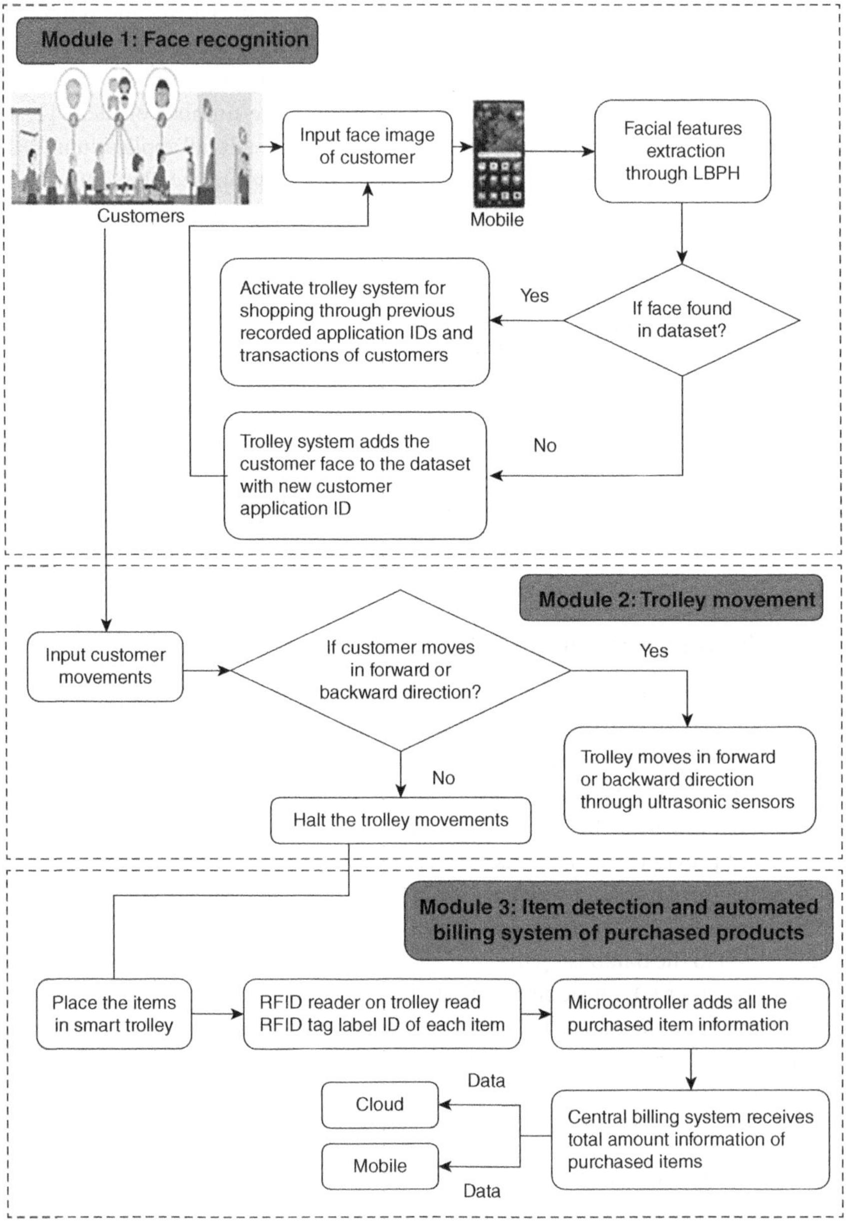

Figure 6.4 Flowchart of proposed trolley architecture. *Source:* Hayesphotography/ Getty Images.

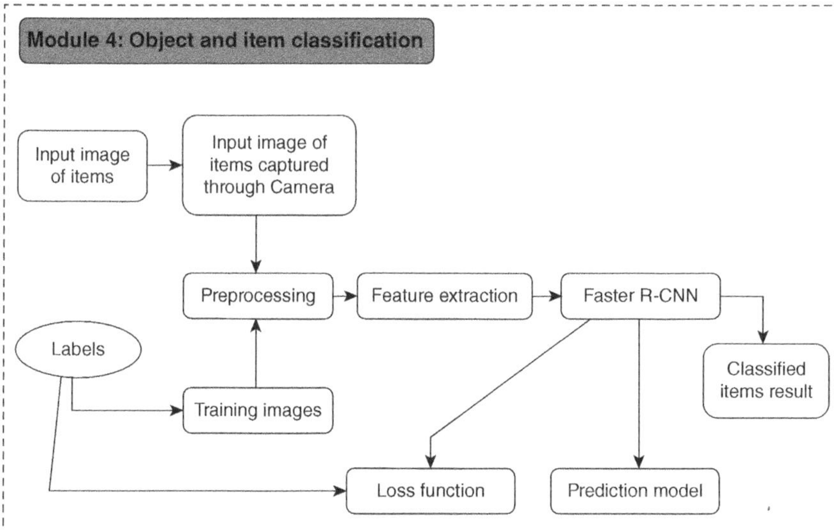

Figure 6.4 (Continued)

technique is used in this process, which converts the input image to a gray-scale image. The mathematical representations for color-to-gray-scale image conversion are given as follows:

i) **Lightness:** $(\max(R, G, B + \min(R, G, B))/3$
ii) **Average:** $(R + G + B)/3$
iii) **Luminosity:** $0.21R + 0.72G + 0.07B$

where R, G, and B represent red, green, and blue components in an image.

e) **Dataset:** To recognize an individual, the dataset includes an alternate number of images of his or her face.
f) **Database:** Face recognition necessarily requires the comparison of face images with a database of images to be identified.
g) **Training Set:** Face recognition is accomplished by matching the face to be identified with a database of recorded face appearances [15]. We include the algorithm faces in the preparation set and identify which individual customers they belong to. When the

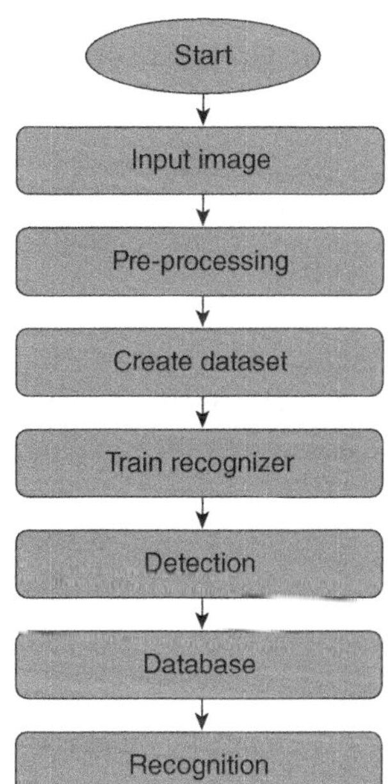

Figure 6.5 Flowchart of the face recognition process.

calculation is required to recognize an unknown face, it uses a training package to do so.

There are three strategies for face recognition: Eigenfaces, Fisherfaces, and LBPH in the OpenCV library. Every one of the three referenced strategies utilizes a preparation set differently. Eigenfaces and Fisherfaces locate a scientific portrayal of the most prevailing highlights of the preparation set as a whole. Each face in the preparation set is independently and widely examined by LBPH. The LBPH algorithm used in this trolley system will not only detect the front faces but can also detect the side faces with 90% accuracy. For face detection of various input images, OpenCV with a Haar cascade classifier is used [16, 17]. The AdaBoost algorithm is used by the Haar cascade classifier to detect multiple face features of different customers, while the LBPH algorithm is used for recognition. The algorithm description for LBPH is as follows:

The LBPH algorithm uses the LBP operator for face extraction and it can be illustrated in Figure 6.6. The formal description of the LBP operator is given in the equation (Eq. [6.1]).

$$\text{LBP}(x_s, y_s) = \sum_{n=0}^{n-1} 2^n S(i_n - i_s)$$

Here (x_s, y_s) are central pixels with intensity i_s; and i_p is the intensity of the neighbor pixel. The sign function (S) is defined as:

$$S(x) = \begin{cases} 1 & \text{if } x \geq 0 \\ 0 & \text{else} \end{cases} \tag{6.1}$$

The idea is to adjust a subjective number of neighbors on a hover with a variable radius, allowing for the capture of neighbor values as shown in Figure 6.7.

For a given Point (x_s, y_s) the position of the neighbor (x_n, y_n), $n \in N$ can be calculated by the equation (Eq. [6.2]).

$$x_n = x_s + R\cos\left(\frac{2\pi n}{N}\right)$$

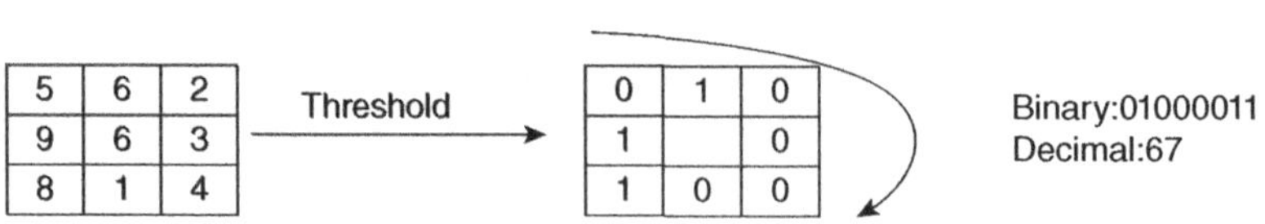

Figure 6.6 Original LBP operator.

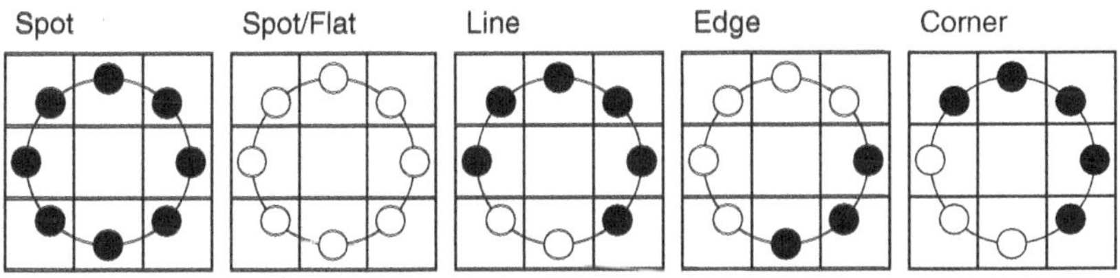

Figure 6.7 Number of neighbors on a hover with a variable radius.

$$y_n = y_s - R\sin\left(\frac{2\pi n}{N}\right), \tag{6.2}$$

where R is the radius of the circle and N is the number of sample points.

This extended LBP operator is also referred to as a circular LBP operator. The point is interpolated if the coordinates of the points on the circle do not correspond to the image coordinates. The bilinear interpolation is implemented in OpenCV (Eq. [6.3])

$$F(x,y) = \begin{bmatrix}1-x & x\end{bmatrix}\begin{bmatrix} f(0,0) & f(0,1) \\ f(1,0) & f(1,1)\end{bmatrix}\begin{bmatrix}1-y \\ y\end{bmatrix}. \tag{6.3}$$

The process flow of the face recognition module using Raspberry Pi Model B with Pi Camera v2 is shown in Figure 6.8.

6.4.2 Module 2 – Trolley Movement

This module comprises sensors for making the trolley an independently moving user-governed vehicle. The sensors have been set on either side of the trolley to act accordingly. The trolley movements are governed by the client's movements if the client is moving forward then the trolley follows while maintaining a certain separation between them automatically. Thus, hauling the check-in products around the store makes the system simple to utilize, and the client can focus on the shopping experience better. The pin schematics of the trolley movement are shown in Figure 6.9.

6.4.3 Module 3 – Items Detection

The client needs to pick the commodity that he/she is keen on and place it in the trolley. The RFID reader channels extract the label ID of each item and add it to the item information when the items are placed in the trolley. The details of the items are then sent to the central billing system of the market and subsequently to

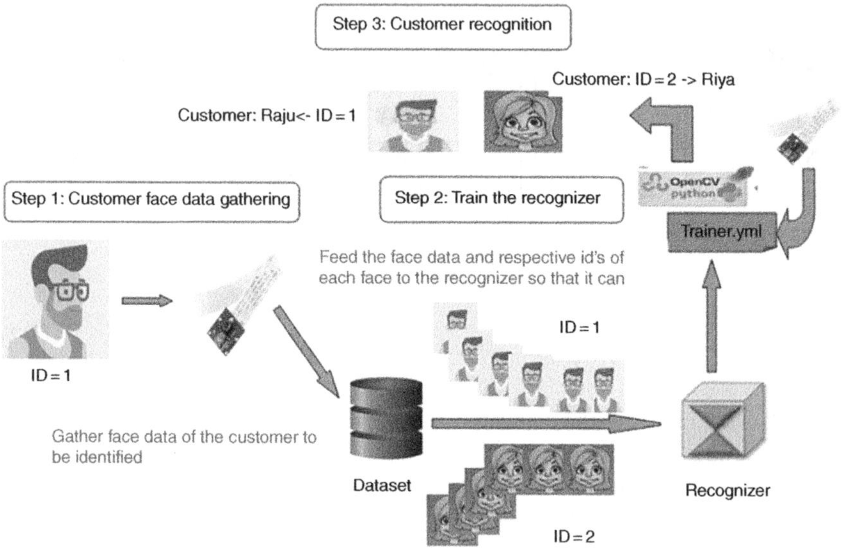

Figure 6.8 Process flow of face recognition module.

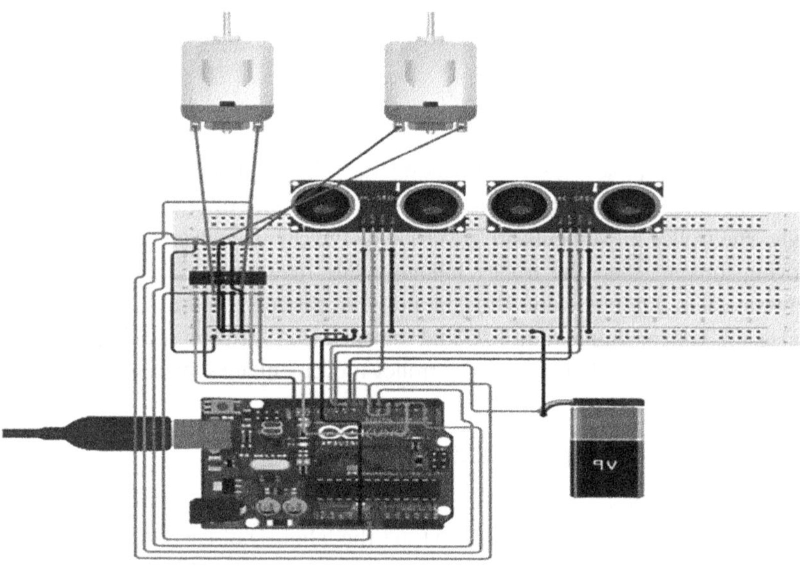

Figure 6.9 Pin schematic of trolley movements.

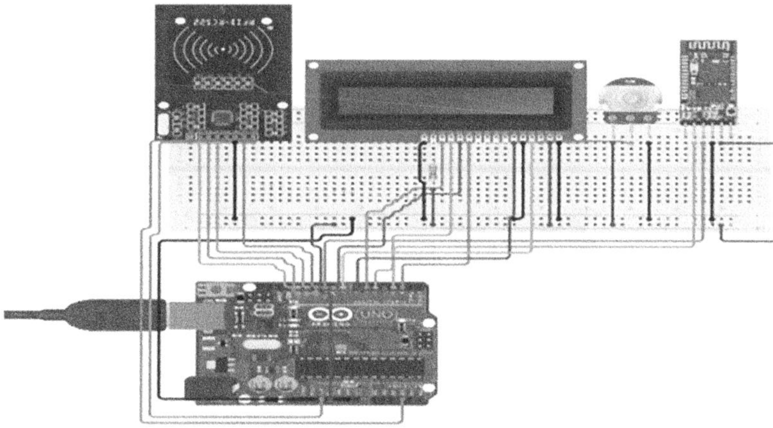

Figure 6.10 Pin schematic of RFID setup.

the cloud and the user's mobile. The aggregate sum will be paid using the client application on the mobile which is linked to the server. These passive tags are scanned more than once to remove the odd chance that there is any label that is not scanned. The client can incorporate the item using the application. In the end, all items are shown in the application as a summary to the client then the payment can be initiated online without the fuss created at the checkout counters. Figure 6.10 shows the pin schematic of the RFID setup.

6.4.4 Module 4 – Object and Item Classification

This module contains a camera that is built into the trolley with numerous assignments and processes attached to it, from face acknowledgment to picture arrangement. The propelled grouping procedures and calculations are utilized to distinguish the items in the stores for better arrangement and for extracting the data utilizing different sorts of a classifier to anticipate the user or customer enthusiasm a better involvement in the store user. The process flow of object classification is as follows: (i) initially the various store images are labeled and trained for analyzing the store surroundings; (ii) real-time images are acquired by the camera, which is then been pre-processed; (iii) feature extraction of the images are done and passed to CNN; and (iv) classified result of output image has been seen with customers present in the particular aisle. Second, it can also anticipate the product details that have been held by the customers in a particular aisle. Figure 6.11 shows the process flow of Faster R-CNN.

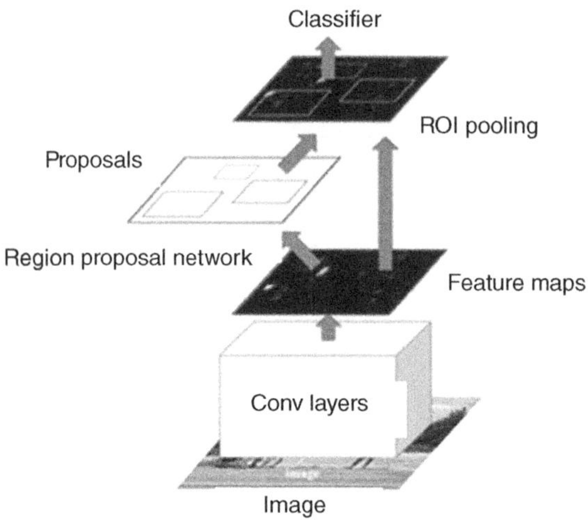

Figure 6.11 Process flow of Faster R-CNN.

6.5 Experimental Results and Discussions

6.5.1 Experimental Results

This section shows the results of each module.

Module 1 – Face Recognition: Datasets are taken by detecting the faces of the customers and each user or customer is given a unique identification number as shown in Figure 6.12. The trainer is trained in a way that each identity of the customer is been replaced by a customer database and recognized. Haar classifier is used to detect the face features, and LBPH is used for recognizing the existing customer, as shown in Figure 6.13.

Module 2 – Trolley Movement: The customer's walking movements are detected by the trolley using ultrasonic sensors connected to the Arduino Uno board. The range from customer and trolley is calculated by sensors and preprocessed to act accordingly. The logic made her is if the person moves forward the trolley moves forward and simultaneously if the customer moves backward the trolley also moves backward as shown in Figure 6.14.

Module 3 – Items Detection: The RFID reader (RC522) connected to the trolley will help it in scanning store products and extracting the product ID with information as well as an aggregate amount for payment as shown in Figure 6.15.

Module 4 – Object and Item Classification: The implementation of the product identification is done by deep CNN algorithms like Faster R-CNN. Various deep-learning Python libraries are used for different purposes in the algorithms. The integrated development environment (i.e. Spyder) is used for the development of the algorithm. Python 3.5 programming language is used for the

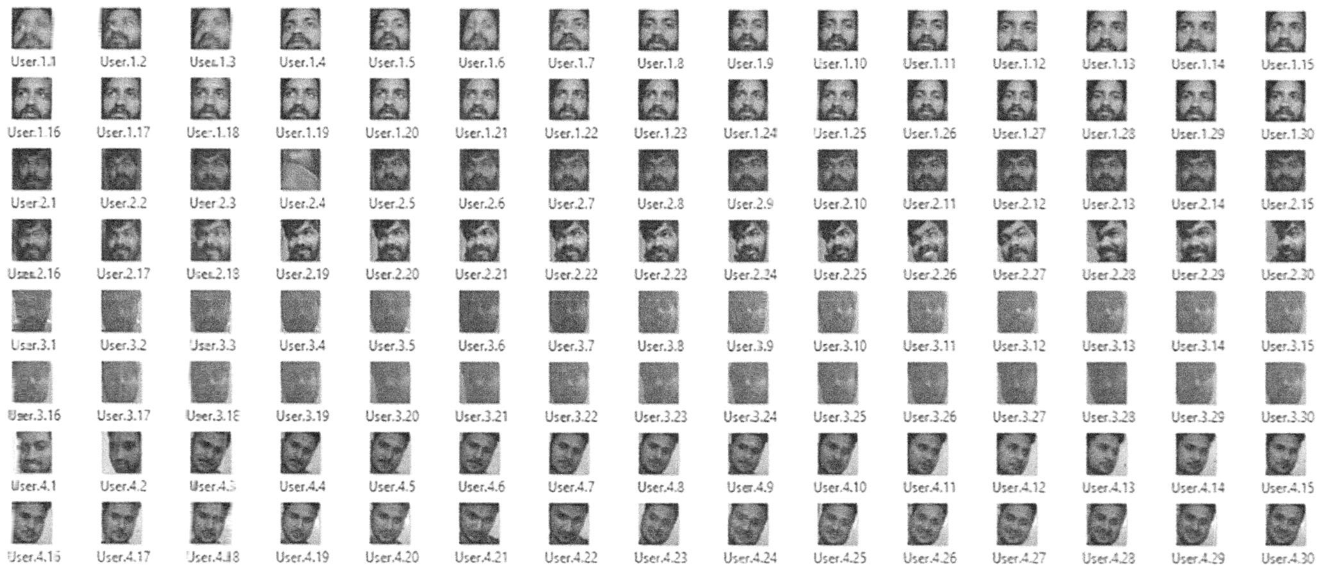

Figure 6.12 Dataset of the customers in-store.

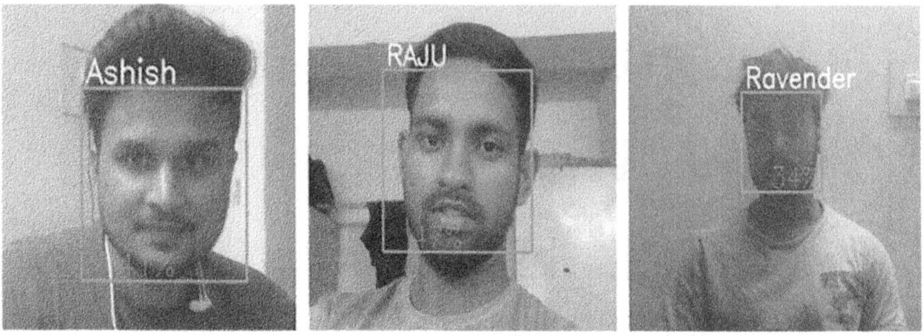

Figure 6.13 Detected customer faces.

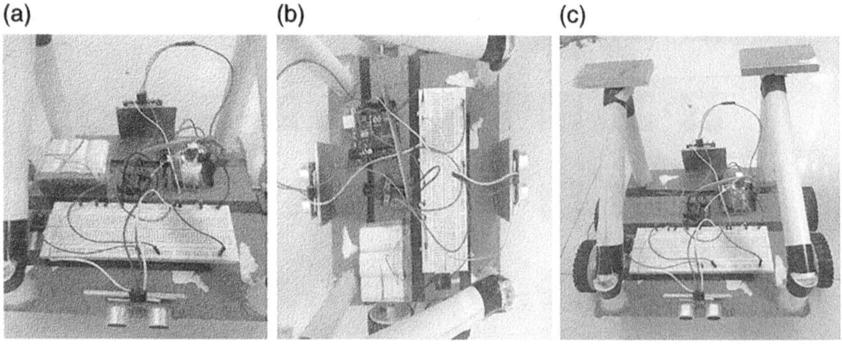

Figure 6.14 Trolley movement's module hardware: (a) forward direction, (b) backward direction, (c) both direction.

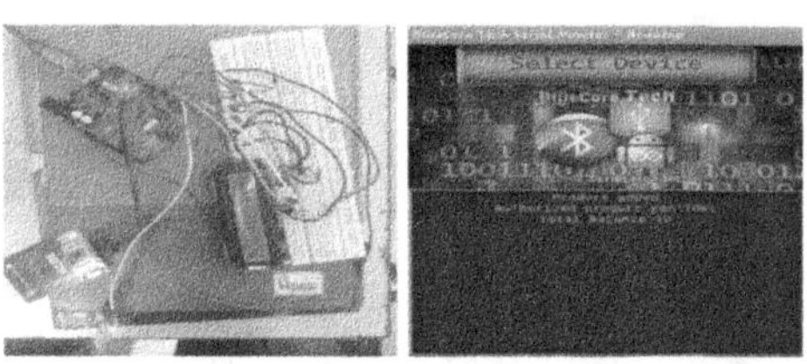

Figure 6.15 RFID module hardware.
Source: Arduino Serial Monitor 2.0.

development of the program. The store item images are gathered and then converted to an XML file using a labeling tool. The labeled images are then given to the CNN for training. The output result after the network is fully trained is shown in Figure 6.16. Training in the simulator is offline training that occurs in an anaconda simulator, which is very useful for debugging and allowing the robot to walk around in an environment without fear of damaging it or its batteries. First and foremost, the initial training images are identified by finding store images in the surrounding area at random. According to the training conditions, it can manage more difficult circumstances in this scenario.

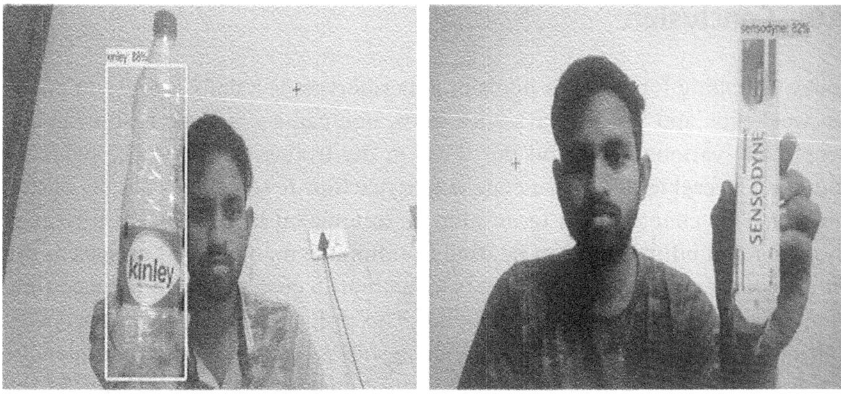

Figure 6.16 Product identification in store.

In a real-time scenario, each trolley has a unique identification number, which will assist the customer so that during the billing, no customer faces the problem.

6.5.2 Discussions on the Security Aspect of IoT

Nowadays, most of the community uses wireless networks, i.e. Wi-Fi, Zigbee, GSM, etc., for automation and live monitoring. But, IoT systems threaten to hack the system. Therefore, the following safety measures are preferable to secure the use of IoT modules in real time [1, 25, 37–40].

a) The users should not share and store personal data and passwords on the cloud. The users use the authenticated website for installing the software.
b) The user should not install any malware without authenticity.
c) Switch off the Wi-Fi network and IoT network when a user does not require it because smartphone-based apps are more vulnerable.
d) IoT service providers should use authenticate software for intelligent device applications.
e) Regular software patches for smartwatches, IoT sensors, IoT gadgets, healthcare apps on smartphones, and other IoT devices should be provided by the IoT service provider. This improves the security of IoT devices. These fixes should be capable of dealing with the most recent malware and viruses.
f) House owners should exercise caution to ensure that no unclaimed IoT gadget is installed or placed on their property without their permission. Later on, hackers may be able to use these devices for malicious purposes.

6.6 Conclusion

Today, shopping for a day in malls or supermarkets is a daily activity in cities. On weekends and special days with sales, one can see a huge rush at malls. People buy various items and put them in the trolley and go to the payment counter. The cashier uses a bar code reader machine to scan the bill at the billing counter. This chapter aims to develop an automated trolley system for smart shopping and billing that can be used to minimize manpower shortages at the billing counter. The main idea here is to facilitate a customer with reduced time spent buying a product in daily shopping. This proposed model is equipped with four modules: (i) face recognition by using LBPH for getting customer details; (ii) user walks independently and prevents collisions with other bodies at the same time by using ultrasonic sensors connected to Arduino Uno board; (iii) Scanning and calculating cart products using RFID tags attached to items and sending data using IoT technology for convenience billing and store alerts on inventory; and (iv) object and item classification for detecting the surrounding store aisle.

References

1 Shraddha, D.D., Trupti, D.J., and Priyanka, S.S. (2016). IoT-based intelligent trolley for the shopping mall. *International Journal of Engineering Development and Research* 4: 1283–1285.

2 Sarac, A., Absi, N., and Peres, S.D. (2010). A literature review on the impact of RFID technologies on supply chain management. *International Journal of Production Economics* 128: 77–95.

3 Ali, Z. and Sonkusare, R. (2014). RFID based smart shopping: an overview. In: *International Conference on Advances in Communication and Computing Technologies*, Mumbai, India (10–11, August 2014). New York: IEEE, pp. 1–4.

4 Yewatkar, A., Inamdar, F., Singh, R., and Bandal, A. (2016). Smart cart with automatic billing, product information, product recommendation using RFID and ZigBee with anti-theft. *7th International Conference on Communication, Computing and Virtualization, Procedia Computer Science,* Mumbai, India (26–27 February 2016). Elsevier, pp. 793–800.

5 Jalkote, V., Patel, A., Gawande, V. et al. (2013). Futuristic trolley for intelligent billing with amalgamation of RFID and ZigBee. *International Journal of Computer Applications Proceedings on International Conference on Recent Trends in Engineering and Technology*: 18–22.

6 Agarwal, A.A., Sultania, S.K., Jaiswal, G., and Jain, P. (2011). RFID based automatic shopping cart. *Control Theory and Informatics* 1: 39–44.

7 McBeath, B. (2013). *The Explosion of Retail Item-level RFID: A Foundation for the Retail Revolution*, 1–8. ChainLink Research Inc.

8 Kumar, R., Gopalakrishna, K., and Ramesha, K. (2013). Intelligent shopping cart. *International Journal of Engineering Science and Innovative Technology* 2: 499–507.

9 Kamble, S., Meshram, S., Thokal, R., and Gakre, R. (2014). Developing a multitasking shopping trolley based on RFID technology. *International Journal of Soft Computing and Engineering* 3: 179–183.

10 Bhatwal, P. and Chamoli, O. (2015). Automated self-billing system using RFID technology. *International Journal of Emerging Trends & Technology in Computer Science* 4: 209–212.

11 Lee, J.G., Kim, M.S., Hwang, T.M., and Kang, S.J. (2016). A mobile robot that can follow and lead humans by detecting user location and behavior with wearable devices. *IEEE International Conference on Consumer Electronics*, Las Vegas, NV (7–11 January 2016). New York: IEEE, pp. 209–210.

12 Chiang, H.H., You, W.T., Lin, S.H. et al. (2016). Development of smart shopping carts with customer-oriented service. *2016 International Conference on System Science and Engineering*, National Chi Nan University, Taiwan (7–9 July 2016). New York: IEEE, pp. 1–2.

13 Wu, B.F., Yao, S. J., Hou, L.W. et al. (2016). Intelligent shopping assistant system. *2016 International Automatic Control Conference*, Taichung, Taiwan (9–11 November 2016). New York: IEEE, pp. 236–241.

14 Ahmed, A., Guo, J., Ali, F., et al. (2018). LBPH based improved face recognition at low resolution. *International Conference on Artificial Intelligence and Big Data*, Chengdu, China (26–28 May 2018). New York: IEEE, pp. 144–147.

15 Zhao, X. and Wei, C. (2017). A real-time face recognition system based on the improved LBPH algorithm. *2nd IEEE International Conference on Signal and Image Processing*, Singapore (4–6 August 2017). New York: IEEE, pp. 72–76.

16 Garg, V. and Garg, K. (2016). Face recognition using Haar cascade classifier. *Journal of Emerging Technologies and Innovative Research.* 3: 140–142.

17 Xiang, Z., Tan, H., and Ye, W. (2018). The excellent properties of dense gird-based HOG features on face recognition compare to Gabor and LBP. *IEEE Access* 6: 29306–29319.

18 Ghai, D., Gera, D., and Jain, N. (2016). A new approach to extract text from images based on DWT and K-means clustering. *International Journal of Computational Intelligence Systems* 9: 900–916.

19 Ghai, D. and Jain, N. (2019). Comparative analysis of multi-scale wavelet decomposition and *k*-means clustering based text extraction. *Wireless Personal Communications* 109: 455–490.

20 Kumar, S., Singh, S., and Kumar, J. (2019). Multiple face detection using hybrid features with SVM classifier. In: *Data and Communication Networks*, Advances in

Intelligent Systems and Computing, 847 (ed. L.C. Jain, V.E. Balas and P. Johri). Singapore: Springer https://doi.org/10.1007/978-981-13-2254-9_23.

21 Kumar, S., Singh, S., and Kumar, J. (2021). Face spoofing detection using improved SegNet architecture with a blur estimation technique. *International Journal of Biometrics* 13: 131–149.

22 Kumar, S., Singh, S., and Kumar, J. (2018). Live detection of face using machine learning with multi-feature method. *Wireless Personal Communications* 103: 2353–2375.

23 Kumar, S., Singh, S., and Kumar, J. (2018). Automatic live facial expression detection using genetic algorithm with haar wavelet features and SVM. *Wireless Personal Communications* 103: 2435–2453.

24 Kumar, S., Prasad, K.M.V.V., Gupta, M.M. et al. (2020). Joyo: the house assistant technology for smart home. *International Conference on Smart Technologies in Computing, Electrical and Electronics*, Bengaluru, India (9–10 October 2020). New York: IEEE, pp. 355–360.

25 Kumar, S., Swetha, S., Kiran, V.T., and Johri, P. (2018). IoT based smart home surveillance and automation. *2018 International Conference on Computing, Power and Communication Technologies*, Greater Noida, India (28–29 September 2018). New York: IEEE, pp. 786–790.

26 Srikrishnaswetha, K., Kumar, S., and Johri, P. (2018). *Comparison study on various face detection techniques. 4th International Conference on Computing Communication and Automation*, Greater Noida, India (14–15 December 2018). New York: IEEE, pp. 1–5.

27 Srikrishnaswetha, K., Kumar, S., and Mahmood, M.R. (2019). A study on smart electronics voting machine using face recognition and aadhar verification with IoT. In: *Innovations in Electronics and Communication Engineering*, Lecture Notes in Networks and Systems, vol. 65 (ed. H. Saini, R. Singh, G. Kumar, et al.). Singapore: Springer https://doi.org/10.1007/978-981-13-3765-9_10.

28 Srikrishnaswetha, K., Kumar, S., and Ghai, D. (2020). Secured electronic voting machine using biometric technique with unique identity number and IoT. In: *Innovations in Electronics and Communication Engineering*, Lecture Notes in Networks and Systems, vol. 107 (ed. H.S. Saini, R.K. Singh, M.T. Beg, and J.S. Sahambi). Singapore: Springer https://doi.org/10.1007/978-981-15-3172-9_31.

29 Sarala, T., Sudha, Y.A., Sindhu, K.V. et al. (2018). Smart electronic trolley for shopping mall. *3rd IEEE International Conference on Recent Trends in Electronics, Information & Communication Technology*, Bangalore, India (18–19 May 2018). New York: IEEE, pp. 2422–2427.

30 Keerthana, S., Inaya, S.A., and Abarna, S. (2018). Design of smart shopping enabler for visually challenged people. *Biomedical & Pharmacology Journal* 11: 2143–2149.

31 Devipriya, D., Sri, V., and Inna, M. (2018). Smart store assistor for visually impaired. *International Conference on Advances in Computing, Communications and Informatics*, Bangalore, India (19–22 September 2018). New York: IEEE, pp. 1038–1045.

32 Gunawan, A.A.S., Stevanus, V., Farley, A. et al. (2019). Development of smart trolley system based on android smartphone sensors. *Procedia Computer Science* 157: 629–637.

33 Anitha, R., Subburam, S., Keerthana, G., and Yoganandarajurs, K.H. (2020). Smart trolley billing system. *European Journal of Molecular & Clinical Medicine* 11: 2264–2270.

34 Das, T.K., Tripathy, A.K., and Srinivasan, K. (2020). A smart trolley for smart shopping. *2020 International Conference on System, Computation, Automation and Networking*, Pondicherry, India (3–4 July 2020). New York: IEEE, pp. 1–5.

35 Shailesh, S., Deb, P.S., Chauhan, R., and Tyagi, V. (2021). Smart trolley. *2021 International Conference on Advance Computing and Innovative Technologies in Engineering*, Greater Noida, India (4–5 March 2021). New York: IEEE, pp. 242–245.

36 Shete, V., Shaikh, A., Sawant, R. et al. (2022). Smart shopping cart for visually impaired individuals. *Review of Computer Engineering Research* 9: 122–134.

37 Patil, S., Mathad, S., Gandhad, S., and Ellemmi, M. (2022). Smart trolley with automatic billing system using Arduino. *IAES International Journal of Artificial Intelligence* 2: 2268.

38 Raju, B.A.N., Ghai, D., and Rawal, K. (2021). Design of futuristic trolley system with comparative analysis of previous models. In: *Big Data Analytics and Intelligent Techniques for Smart Cities* (ed. K.B. Prakash, J. Nayak, and B. Madhhav), 223–247. CRC Press.

39 Ghai, D., Tiwari, S., and Das, N.N. (2021). Bottom-boosting differential evolution based digital image security analysis. *Journal of Information Security and Applications* 61: 1–8.

40 Nanda, S.K., Ghai, D., and Pande, S. (2021). VGG-16-based framework for identification of facemask using video forensics. In: *Proceedings of Data Analytics and Management*, Lecture Notes on Data Engineering and Communications Technologies, vol. 91 (ed. D. Gupta, Z. Polkowski, A. Khanna, et al.). Singapore: Springer. https://doi.org/10.1007/978-981-16-6285-0_54.

7

Enhancing Stock Market Predictions Through Predictive Analytics

Ameya Patil[1], *Shantanu Saha*[2], *and Rajeev Sengupta*[3]

[1] *Faculty of School of Business, MIT World Peace University, Pune, India*
[2] *Faculty of Ramcharan School of Leadership, MIT World Peace University, Pune, India*
[3] *Faculty of School of Business, Dr. Vishwanath Karad MIT World Peace University, Pune, India*

7.1 Introduction

The stock market is a gathering of buyers and sellers and provides a venue for trading the shares of various corporations. Financial markets assist economic growth by facilitating international trade, gathering, discounting, and disseminating information about organizations' and economies' future prospects. Financial securities amounting to trillions of dollars are traded daily on the stock exchanges around the world. With the advent of technology and popularization of online trading applications, such as Zerodha and Upstox providing convenience, interest of common people has also grown in stock markets. In emerging countries like India, many were attracted to stock markets during the Covid-19 pandemic. This was evident by the record number of new demat accounts opened on online platforms such as Zerodha in 2020. Stock market forecasting has long been a difficult and perplexing issue, which has caught the attention of the general public and, the professionals in this area, and of experts like economists and computer scientists. Successfully anticipating stock price movements can result in alluring rewards through strategic action, making stock market trend prediction a task worthy of significant focus and effort. Stock market forecasting is a very difficult undertaking, and various aspects must be addressed to make effective market predictions. Considering the fact that stock price is dependent on numerous aspects, only taking into account one or two criteria will not be sufficient to provide adequate information to estimate the stock price movement. Hence, it is appropriate to include

Intelligent Techniques for Predictive Data Analytics, First Edition.
Edited by Neha Singh, Shilpi Birla, Mohd Dilshad Ansari, and Neeraj Kumar Shukla.

Published 2024 by John Wiley & Sons, Inc.

as much pertinent and helpful data as possible to ensure a more accurate estimate. Profits are realized by those investors who make the right buy and sell choices. The stock market prediction aimed at analyzing the trends is convoluted because the stock market is inherently noisy and has a lot of volatility when it comes to trends [1]. Several strategies have been developed in existing methodologies to predict stock market trends. The reliability of financial market forecasting and the viability of lucrative model-based trading are heavily impacted by a gamut of factors. These include the market's maturity, the forecasting method utilized, the time frame for which forecasts are made, as well as the methodology employed to evaluate the model and simulate trade [2].

The efforts to stock market prediction begin with fundamental analysis and technical analysis. Fundamental analysis is a sort of investment analysis wherein the company's past and estimated sales, profits, and other financial parameters are analyzed along with other economic based on factors that affect the value of the company. This type of analysis assesses the stock price in light of its intrinsic value, viz., fair value. This approach is best suited for long-term forecasting. Technical analysis is based on charts and trends. This analysis makes use of historical stock prices to forecast price trends. The input features for machine learning (ML) and AI models can also be employed with experience-based technical indicators. The fundamental idea of ML is to draw knowledge through data. Next, time series forecasting using linear models for stock market forecasting is in vogue, including ARIMA and GARCH. The 21st century was defined by the rapid development of electronic markets and the emergence of computer-driven high-frequency trading enabled by computing technologies [3]. Nonlinear models such as artificial neural networks (ANNs) have then been used to overcome the limitations of linear models, considering the chaotic nature of stock markets. Utilizing time series analysis technology for stock forecasting is vital for creating stock trading strategies and earning greater trading returns [4]. ML techniques for predicting the stock market are in vogue. Numerous studies that utilized ML for financial time series forecasting appear to be able to predict future market events with up to 80% accuracy [2]. Analysts and investors place a premium on the ability to anticipate shifts in stock prices. The challenge of making accurate and efficient stock market predictions is quite complex, and various elements should be taken into account. Stock prices respond to market changes, quarterly earnings announcements, and evolving consumer patterns. Traders utilize a variety of technical indicators derived from daily stock data.

The purpose of this chapter is to provide a thorough examination of several stock prediction approaches for forecasting future trends and stock returns. This chapter takes into account a number of stock market forecasting methods that have been used in previous research studies.

7.2 Factors Influencing Stock Prices

Earnings in the stock market can be forecasted using traditional methods based on past stock returns, as well as financial and economic indicators. The forecast of stock market returns led investors to investigate the grounds of predictability. The variety of variables that can be used to make forecasts should be taken into consideration [5]. Forecasting stock trends is a difficult technique since it is dependent on a variety of factors, such as the specific market patterns, expectations of investors and traders, and their financial circumstances. Several news and events also impact stock values. Furthermore, the stock price list is by definition dynamic, complicated, noisy, nonparametric, and nonlinear. Financial time series forecasting is challenging due to a variety of complicated properties such as volatility, anomalies, noise, and shifting trends. Enormous amounts of data regarding stocks are generated on a daily basis and are accessible from various channels such as stock exchange websites, economic reviews, company reports. Different research studies utilized distinct sets of factors as input to their prediction models, as no consensus regarding all aspects crucial to stock market forecasting exists [6].

The following data have been so far used to predict stock prices:

1) **Market Data:** Market data refers to the collection of information and statistics related to financial markets and the trading activities that take place within them. It includes various types of data that provide insights into the prices and volumes. Also included are the current market price, opening price, closing price, high and low prices for a given time period.
2) **Text Data:** Individually written text from sources, such as news, social media, and web searches is referred to as text data. As an alternate data source, these statistics are difficult to acquire and evaluate. However, they can offer valuable insights that are not reflected in aggregate market data. However, they may provide helpful information that is not reflected in aggregate market data. A sentiment component can be added to text data through the use of sentiment analysis (such as positive, negative, or neural). The sentiment component is then applied to prediction.
3) **Macroeconomics Numbers:** Macroeconomic indicators, such as the gross domestic product (GDP), inflation rates, interest rates, reveal the economic realities of a particular nation, area, or sector. These indicators have some connection to the stock market because they reflect the market as a whole and can verify the validity of a stock price increase or decrease.
4) **Knowledge Graph Data:** Different companies and markets interact in some way. For instance, similar news may influence the movement of stocks in the same industry. Knowledge graph data from open sources can be leveraged to enhance prediction accuracy utilizing graph neural networks.

5) **Image Data:** Image data can be used to create visualizations of stock market data, such as candlestick and line charts. Some researchers have investigated using satellite imagery or aerial images to gather information related to economic activity, such as analyzing parking lot occupancy rates or estimating oil storage levels, which could indirectly impact certain stocks or sectors.
6) **Fundamental Data:** Accounting data, such as assets and liabilities, is the most typical sort of fundamental data, which is reported quarterly. Financial information of a corporation, such as its sales, earnings, assets, liabilities, and other financial ratios, comprise fundamental data. Fundamental data reveals a company's financial health and can influence its stock price.
7) **Recommendation Data:** Analyst recommendation data refers to information provided by financial analysts or brokerage companies regarding their opinions and recommendations regarding particular stocks or investment opportunities. Analysts do research and analysis on companies and industries, and issue reports including buy, sell, or hold recommendations for specific stocks based on their findings.

7.3 Can Markets Be Predicted?

The efficient markets hypothesis (EMH), a central tenet of finance theories, holds that market price represents all available information. As a result, risk-free profits cannot be achieved [7]. According to EMH, as market prices vary in a random manner, future values cannot be predicted using data from the past. EMH may be broken if an investor can consistently outperform the market in terms of risk-adjusted return. Despite being prevalent, this hypothesis is not widely accepted since it cannot account for a number of minor but major instances of market inefficiencies. These anomalies include higher returns posted by institutional investors, the market's over-reaction and under-reaction, and high volatility of asset prices.

7.3.1 Different Stock Prediction Techniques

Many forecast models presume that the past may be utilized to envisage future stock market movements. Several ML and econometric methods have been used in stock market forecasting. Financial time series forecasting, frequent use of ML technologies, claims that sophisticated forecasting algorithms effectively predict price movements in stock markets, and trading on these predictions can generate a profit. Financial economists, on the contrary, refer to the informational efficiency of financial markets, which calls into question price prediction and prospects for lucrative trading [2]. Several research studies have utilized a combination of two or more methods collectively to forecast the stock markets.

A combination of fundamental data and technical indicators has been used in many research works.

Fundamental Analysis: In this type of analysis, underlying elements that affect businesses or sectors are used as predictive attributes [8]. Fundamental analysis focuses on fundamental information, mainly contained in the annual reports of companies, such as a company's sales and profits, the annual growth rate in various financial parameters, company's position in the market and its future prospects [9]. One of the important aspects of fundamental analysis is economic analysis, which focusses on the macroeconomic parameters such as national income, exports data and interest rates which ought to impact the operations of companies listed on a stock exchange. In addition, sectoral information, which ought to impact stocks belonging to specific sectors, is also incorporated into fundamental analysis. Using this information, the company's future cash flows and thereby, the company's value is arrived at, and is used to forecast its stock price. Early studies on stock market prediction have shown an interest in fundamental indicators and related variables [10, 11]. Fundamental analysis, when incorporated with ANNs can result in superior returns [12, 13].

Technical Analysis: Technical analysis refers to assessment of previous stock price and volume information to forecast stock price changes [14]. Technical analysis is based on analyzing the current demand–supply of the tradable instrument. A variety of technical indicators are used to generate "buy" and "sell" signals. The majority of research works regarding stock price prediction through ML techniques have employed technical indicators [15]. Technical analysis and historical market movements have the ability to anticipate daily returns in an active trading environment [16]. Technical indicators like moving averages, oscillators, and trend lines are used. Several research studies have employed technical indicators as input for forecasting stock markets [17, 18]. In the age of AI, technical indicators have been provided as inputs to a neural network to improve prediction accuracy [19, 20].

Sentiment Analysis: With the easy availability of the Internet globally, internet and social media disperse huge amount of information related to stocks as well as facilitate users to engage in expressing their opinion related to the same. The sentiment analysis approach involves analyzing investor sentiment and market psychology to gauge market direction. Sentiment analysis includes monitoring news, social media, and market indicators to assess the overall mood and sentiment of market participants. Incorporating certain public mood dimensions can improve the prediction accuracy regarding stock markets [21]. When applied correctly, investments assisted by sentiment analysis tools can yield rewards. Together, sentiment analysis and collective intelligence can automatically compute the stock market's future performance [22]. Feature extraction of social media text data can describe the emotional tendencies of investors and enhance the precision of stock price forecasting [23]. The addition of news and sentiments can improve deep learning models' performance as regards to accuracy in price prediction. Textual information in news headlines is related to stock price forecasting [24, 25] used technical indicators from stock prices and sentiments from textual news articles to build a layered deep learning neural network model prediction model, which brings more prediction enhancements.

7.4 Using Predictive Analytics for Stock Prediction

A data-driven methodology called predictive analytics combines statistical algorithms and historical data to forecast upcoming events or results. Predictive analytics is the process of creating algorithms that can anticipate future stock prices or locate prospective trading opportunities using previous stock market data. Enormous amount of structured and unstructured heterogeneous data is generated related to stock markets and the individual scrips therein, can be effectively analyzed through advanced ANNs. ML and AI are extensively used when it comes to predictive analytics.

Here are some typical steps in the stock prediction process utilizing predictive analytics:

1) **Data Collection:** Collect historical data for the stock(s) you wish to examine. This often comprises daily or intraday pricing data, trade volumes, and other pertinent information such as financial statements or economic indicators. In addition, for a more comprehensive study, company-specific data and news sentiment data can be included.
2) **Data Preprocessing:** Clean and preprocess the data by dealing with missing values, outliers, and, if necessary, normalizing the data. This phase ensures that the data is in an appropriate format for analysis.
3) **Feature Selection:** Determine the most important traits or elements that may have an impact on stock pricing. Technical indicators, financial ratios, economic indicators, and news mood ratings are examples. Regarding feature selection, some researchers investigate the relationship between stock prices and certain novel variables, such as political considerations, macroeconomic factors, and investor sentiment, among others, and include them in the prediction model.
4) **Model Selection:** Based on your facts and goals, select a suitable predictive model. Regression methods, time series models such as ARIMA, GARCH, or ML algorithms, or neural networks are generally used for stock prediction.

5) **Training the Model:** Divide the historical data into two sets: training and testing. Use a training set for training the predictive model by fitting it to the historical data, adjusting model parameters, and optimizing performance.
6) **Evaluation and Tuning:** Using the testing set, evaluate the predictive model's performance. Mean squared error (MSE), root mean squared error (RMSE), accuracy, and precision-recall measurements are common stock prediction evaluation metrics. If the model's performance is poor, hyperparameter tuning and model tweaks can be made to increase accuracy.
7) **Prediction and Decision Making:** Once the model has been trained and evaluated, use it to forecast future stock prices or find potential trading opportunities based on new data. Monitor and analyze the expected results to determine the model's effectiveness and, if necessary, adjust the method.

7.4.1 Stock Prediction Using ML and AI

Experts have long stressed the need for prediction, especially as the environment has gotten more chaotic and turbulent. Information technology advancements for forecasting can be aided by information about the necessary set of indicators. Forecasts, on the contrary, are based on previous events and cannot easily incorporate all essential information, particularly unstructured information. The combination of AI and data analytics in forecasting would surely benefit investors.

Because of the irregularity of the financial market, simplistic models cannot accurately forecast future asset values [26]. Hence, some investors and institutions use ML and AI algorithms to find patterns or trends in massive amounts of data. These techniques can help in generating predictions or identifying potential trading opportunities based on historical data. Many ML models have shown the value of a novel ML model for time series forecasting utilizing financial market data such as stock prices, currency rates, or financial indices. They have also identified profitable trading opportunities based on model forecasts. It has been observed that ML algorithms perform better than most traditional stochastic methods for predicting financial markets [27].

7.5 Neural Networks

Neural networks, as a wise information mining approach and profound learning methods, are useful in learning complicated forms of information by applying regulated learning models, where it gradually learns from datasets and experience. Neural networks provide a revolutionary technique that does not necessitate pre-specification during the modeling phase because they understand the link inherent in the variables on their own. This is particularly beneficial in the security investing arena, where much is assumed, but little is understood about the processes that determine stock prices. Due to their aptitude for learning nonlinear mappings [28] between input and output, neural networks are used to forecast stock market prices [29]. Because of the large amounts of information generated in capital markets, machines would master many designs, allowing them to make reasonable forecasts [30]. Neural networks have seen widespread use for stock prediction owing to their capability to learn intricate patterns and relationships in data.

7.5.1 Artificial Neural Network

ANN is an algorithm meant to comprehend complex problems that cannot be addressed by standard ML methods or accessible neural networks. ANNs have transferred human learning ability from their experiences into a computer environment that can mimic the working principles of the human brain. ANNs are constructed with a more intricate network of interconnections than the human brain [31]. The approach uses algebraic equations to direct data to a model or time series line. By means of a self-learning approach, ANN can discover hidden features. These are good approximators that can find input and output relationships of a huge complex dataset. Hence, ANN is a suitable alternative for predicting stock prices. ANN is utilized to reveal the complex correlations between stock performance and related financial and technical variables [32]. Typically, financial variables serve as the inputs and stock performance is the resultant output, for ANN applications. Numerous research studies have revealed the existence of nonlinearity in financial markets and the applicability of ANN to uncover this relationship. ANN has been widely utilized to achieve accurate stock market forecasting [1, 32]. Several research works have revealed that ANNs outperformed statistical models like ARIMA [33]. At the same time, the combination of statistics and learning models has polished neural networks. The input variables and network parameters for ANNs must be carefully chosen. The ability to accurately predict the future is highly dependent on learning algorithm employed during ANN training [34, 35] found FNN and RNN types, such as LSTM, GRU, and CNN, quite effective in predicting the riskiness of Indian stock market indicated by the fear factor called volatility index (VIX).

ANNs are neural networks with one or more layers that are fully interconnected. A sample ANN comprising input layer, output layer, and three hidden layers is shown below (Figure 7.1). Every node in a layer is connected to each node in the layer above it. It is feasible to make the network deeper by increasing hidden layers.

The backpropagation (BP) algorithm is a widely used optimization technique to train ANNs. BP is a supervised learning algorithm used to adjust the weights and biases of neural networks during the training process, aiming to minimize the

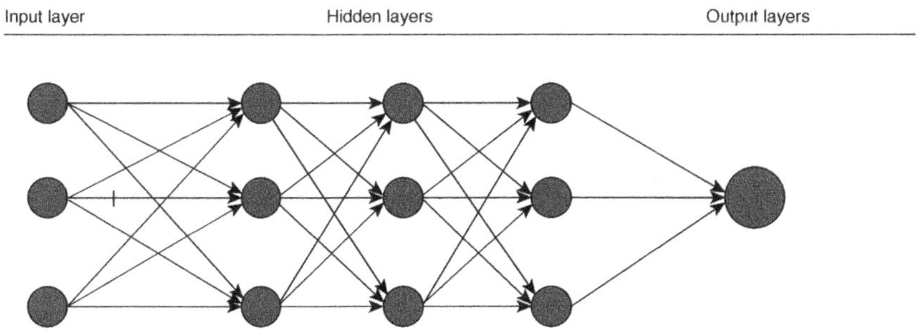

Figure 7.1 A sample of artificial neural network.

difference between predicted and actual outputs. In majority of the related researches, BP algorithm has been used widely, especially for training the data set. However, owing to the high levels of noise and complicated dimensionality in stock market data, ANN has some constraints when learning patterns. Hence, many research studies have applied hybrid models, wherein the traditional BP algorithm has been used in conjunction with other algorithms. Hybrid models of ANN and genetic algorithms (GAs) for feature discretization have been widely used in this context. The classical BP learning algorithm, when aided by global search techniques such as GA and simulated annealing (SA), can significantly increase prediction accuracy [36]. Modifications can be made to neural networks using enhanced algorithms such as the improved sparrow algorithm. Utilizing it for tuning initial weight as well as threshold of BP neural network resulted in a highly accurate short-term forecast of stock values [37]. Improved sine cosine algorithm (ISCA) combined with BP neural networks, along with Google Trends, achieved prediction accuracy to the tune of 87% [38]. Neural networks when assisted using nonlinear independent component analysis (NLICA) for feature extraction improves the prediction accuracy [39].

7.5.1.1 Recurrent Neural Network (RNN)

Depending upon their architecture and purpose, there are numerous ANN kinds. Feedforward neural network (FNN) is a basic type of ANN, where information flows unidirectionally, viz., from the input layer to the output layer. FNN's take historical stock price information, along with other pertinent features such as trading volume, technical indicators, or economic data, as input and predict future stock prices or market trends. Stock market prediction studies have mainly used RNN, a neural network architecture designed to handle sequential data by considering the temporal aspect of the information. The main characteristic of RNN is its knack to capture and propagate information from previous time steps to current time step, allowing them to model dependencies and patterns in sequential data. When it comes to stock prediction, RNNs can be used to process historical stock price data and other relevant features over time, capturing patterns and relationships that can aid in forecasting stock prices.

The RNN architecture known as long short-term memory (LSTM) is particularly good at identifying long-term dependencies and patterns in sequential data. LSTM networks consist of memory cells that store and update information over time, as well as gates that regulate the flow of information through the network. The three gates that make up an LSTM are input gate, forgetting gate, and output gate. Information not complying with the algorithm will be erased by the forgetting gate. Hence, inly the compliant data remains [40]. For the purposes of forecasting stock prices, LSTM networks can be trained to learn patterns and relationships in historical stock price data, as well as other relevant features. By considering the sequential nature of the data and incorporating memory cells, LSTM networks can potentially capture trends, seasonality, and other patterns that are significant for predicting future stock prices.

LSTM basically operates as a three-step process. In the first step, the LSTM is to decide which information to be omitted from the cell in that particular time step. This decision is made with the help of the sigmoid function, which computes the function using the prior state and the current input. The second layer contains two functions, viz., the sigmoid function and the tanh function. The sigmoid function determines which values are allowed to pass (0 or 1). Tanh function assigns weightage to the values passed, deciding their level of importance from −1 to 1. The third stage determines the ultimate result. First, a sigmoid layer needs to be run, which determines what parts of the cell state make it to the output. The cell state is then passed to the tanh function, which pushes values between −1 and 1 and multiplies it by sigmoid gate output.

Convolutional neural networks (CNNs) can be applied to certain aspects of stock prediction. CNN is a feed-forward neural network. More hidden layers are present in CNN vis-a-vis traditional neural networks. Designing CNN's architecture for stock prediction challenges is conceivably a key problem that merits additional research. Primarily, CNN can be employed for data visualization purposes, converting financial data into visual representations such as candlestick charts, market depth plots, or heat maps. Such visualizations can help traders and analysts gain insights into market trends and patterns. Second, CNNs can potentially be used for feature extraction and preprocessing of financial time series data. Next, CNN can also be used in sentiment analysis by processing textual data such as news articles, social media posts, or financial reports related to the stock market. CNN can also analyze alternative data sources, such as satellite imagery, web-scraped data, or visual data from sensors. A deep CNN-based framework applied to extract higher-level features from a large number of financial variables can significantly improve prediction accuracy [5].

RNNs are hampered by the vanishing gradient problem [41], which makes it difficult for them to recognize long-term dependencies in sequential data. Another RNN called gated recurrent unit (GRU) is employed to solve this problem. The primary characteristic of a GRU is the incorporation of gating mechanisms that allow the network to control the flow of information across time, hence enabling the network to retain and update pertinent information over longer sequences. By means of gating mechanisms, GRUs excel in capturing long-range dependencies in sequential data, making them better suited for applications where long-term information retention is critical. Unlike the LSTM, where the output gate controls what content is seen by the other units, the GRU does not have controlled memory exposure. GRUs can be used for time series forecasting, where historical stock prices and other pertinent information are fed into the network, and the model predicts future stock prices or market trends. GRUs' capacity to detect temporal dependencies in the data may be useful for spotting patterns and trends in stock price movements. By learning meaningful representations of input data, the model can capture complex patterns that may impact stock prices. In addition, GRUs can extract features from time series data. The time series events in input may be successfully captured by LSTM and GRU models. In addition, a fully connected neural network is utilized to combine numerous independent prediction outcomes to increase prediction accuracy [42]. GRUs can be employed for sentiment analysis using textual data like financial news articles or social media posts related to the stock market. GRU model fed with historical prices and the sentiment score delivered great prediction results [24]. Incorporating financial news sentiments along with stock attributes as the input can considerably improve the performance of LSTM and GRU [43]. Deep learning models which integrate multiple several CNN pipelines along with unidirectional or bidirectional GRUs have been instrumental in enhancing stock prediction [44]. One of the variants, nested GRU (NGRU) employs a nested structure to boost model depth and eliminate noise from stock data features. Learning the features of stock data, NGRU has been found to be more sensitive and effective.

7.5.1.2 Deep Learning Models

Deep learning is focused on building multilayered ANNs. Deep learning approaches use data to tackle complex problems with fewer assumptions. Deep learning automatically extracts features from data, requiring little to no human involvement in feature selection. The term "deep" alludes to the numerous layers that make up the neural network architecture. Deep learning employs neural networks with multiple hidden layers, allowing the network to learn representations of the data at different levels of abstraction. Each layer in the network extracts higher-level features from the representations learned by the previous layers, thus creating a hierarchy of learned features. Deep neural networks integrate benefits of deep learning and neural networks. Hence, they can be used to address nonlinear problems more effectively than traditional ML algorithms [45]. Such networks yield greater precision with less human effort and minimal tuning. Deep learning has been applied to stock prediction, particularly in the context of time series analysis, because of its ability to understand complicated patterns and representations from past data. Deep learning models need massive input data to build a complicated neural network model [46]. Here, market data is the best choice and is the most popular. Text data is the second most used type. This is because social media and online news sites are so popular, and it is easy to get this type of data. The majority of deep learning systems produce forecasts based on previous stock price, volume, or derivative data [4]. Deep learning may be appealing for stock market prediction at high frequencies since it can extract characteristics from a large amount of raw data without relying on prior knowledge of predictors [47]. Deep learning methods come in a wide variety of network structures, activation functions, and other model parameters, and it is well-recognized that the way data is represented has a significant impact on how well these algorithms perform. According to empirical findings, deep neural networks can improve prediction accuracy by extracting additional data from autoregressive model residuals.

7.6 Conclusion

Stock market forecasting has been an area of great interest to the practitioners as well as the investors. However, this task is difficult since stock markets are complex, volatile, and nonlinear [41]. The early studies in stock market forecasts were mainly in fundamental analysis [10] and technical analysis [17]. With the advent of AI, the various variables used in technical and fundamental analysis are feed as inputs to the ANNs [4, 20]. Such studies have been widely used in stock market forecast, along with time series forecasting. Also, news articles and sentiment analysis data are being incorporated into neural networks for stock price prediction [45]. ANNs have also seen development in terms of improvements in the existing technology. Several research studies have used hybrid models for stock forecasting. The recurrent neural networks especially LSTM, GRU, and CNN are now being widely used along with some permutations and combination of these networks. The traditional back propagation algorithm used in neural networks has been added by several other algorithms to improve prediction accuracies. Several research studies have fine-tuned and developed ANN to improve prediction accuracy, and the efforts in this direction are in full swing. The accuracy of predictions can be increased by combining several approaches, employing boosting algorithms, and including feature selection. Stock prediction will always be an area of research and will be defined by the fundamental data and technical indicators, financial news and sentiments, the advent of technologies such as hybrid ANNs, as well as the algorithms used therein for training and prediction purposes. The future stock prediction techniques should take into account a myriad factor affecting stock markets, including sentiments, political news, government policies, and others. Hence, the stock prediction models developed in the future should incorporate data from different sources. Furthermore, ongoing advances in artificial intelligence and ML algorithms should be leveraged to create models that are adaptive, capable of learning from new data and better equipped to deal with unexpected market shifts.

References

1 Gandhmal, D.P. and Kumar, K. (2019). Systematic analysis and review of stock market prediction techniques. *Computer Science Review* 34: 100190. https://doi.org/10.1016/j.cosrev.2019.08.001.

2 Hsu, M.W., Lessmann, S., Sung, M.C. et al. (2016). Bridging the divide in financial market forecasting: machine learners vs. financial economists. *Expert Systems with Applications* 61: 215–234. https://doi.org/10.1016/j.eswa.2016.05.033.

3 Lehoczky, J. and Schervish, M. (2018). Overview and history of statistics for equity markets. *Annual Review of Statistics and Its Application* 5: 265–288.

4 Liu, H., Zhao, T., Wang, S., and Li, X. (2023). A stock rank prediction method combining industry attributes and price data of stocks. *Information Processing & Management* 60 (4): 103358. https://doi.org/10.1016/j.ipm.2023.103358.

5 Hoseinzade, E. and Haratizadeh, S. (2019). CNNpred: CNN-based stock market prediction using a diverse set of variables. *Expert Systems with Applications* 129: 273–285. https://doi.org/10.1016/j.eswa.2019.03.029.

6 Kumbure, M.M., Lohrmann, C., Luukka, P., and Porras, J. (2022). Machine learning techniques and data for stock market forecasting: a literature review. *Expert Systems with Applications* 197: 116659. https://doi.org/10.1016/j.eswa.2022.116659.

7 Fama, E.F. (1970). Efficient capital markets: a review of theory and empirical work. *The Journal of Finance* 25 (2): 383–417. https://doi.org/10.2307/2325486.

8 Bustos, O. and Pomares-Quimbaya, A. (2020). Stock market movement forecast: a systematic review. *Expert Systems with Applications* 156: 113464. https://doi.org/10.1016/j.eswa.2020.113464.

9 Nair, B.B. and Mohandas, V.P. (2015). An intelligent recommender system for stock trading. *Intelligent Decision Technologies* 9 (3): 243–269. https://doi.org/10.3233/IDT-140220.

10 Ou, J.A. and Penman, S.H. (1989). Financial statement analysis and the prediction of stock returns. *Journal of Accounting and Economics* 11 (4): 295–329. https://doi.org/10.1016/0165-4101(89)90017-7.

11 Holthausen, R.W. and Larcker, D.F. (1992). The prediction of stock returns using financial statement information. *Journal of Accounting and Economics* 15 (2–3): 373–411. https://doi.org/10.1016/0165-4101(92)90025-W.

12 Lam, M. (2004). Neural network techniques for financial performance prediction: integrating fundamental and technical analysis. *Decision Support Systems* 37 (4): 567–581. https://doi.org/10.1016/S0167-9236(03)00088-5.

13 Shen, K.Y. and Tzeng, G.H. (2015). Combined soft computing model for value stock selection based on fundamental analysis. *Applied Soft Computing* 37: 142–155. https://doi.org/10.1016/j.asoc.2015.07.030.

14 Lohrmann, C. and Luukka, P. (2019). Classification of intraday S&P500 returns with a Random Forest. *International Journal of Forecasting* 35 (1): 390–407. https://doi.org/10.1016/j.ijforecast.2018.08.004.

15 Kumar, G., Jain, S., and Singh, U.P. (2020). Stock market forecasting using computational intelligence: a survey. *Archives of Computational Methods in Engineering* 28 (3): 1069–1101. https://doi.org/10.1007/s11831-020-09413-5.

16 Bitvai, Z. and Cohn, T. (2015). Day trading profit maximization with multi-task learning and technical analysis. *Machine Learning* 101: 187–209. https://doi.org/10.1007/s10994-014-5480-x.

17 Brock, W., Lakonishok, J., and LeBaron, B. (1992). Simple technical trading rules and the stochastic properties of stock returns. *The Journal of Finance* 47 (5): 1731–1764. https://doi.org/10.1111/j.1540-6261.1992.tb04681.x.

18 Bessembinder, H. and Chan, K. (1998). Market efficiency and the returns to technical analysis. *Financial Management* 5–17. https://doi.org/10.2307/3666289.

19 Thawornwong, S., Enke, D., and Dagli, C. (2003). Neural networks as a decision maker for stock trading: a technical analysis approach. *International Journal of Smart Engineering System Design* 5 (4): 313–325. https://doi.org/10.1080/10255810390245627.

20 Sezer, O.B., Ozbayoglu, M., and Dogdu, E. (2017). A deep neural-network based stock trading system based on evolutionary optimized technical analysis parameters. *Procedia Computer Science* 114: 473–480. https://doi.org/10.1016/j.procs.2017.09.031.

21 Bollen, J., Mao, H., and Zeng, X. (2011). Twitter mood predicts the stock market. *Journal of Computational Science* 2 (1): 1–8. https://doi.org/10.1016/j.jocs.2010.12.007.

22 Devi, K.N. and Bhaskaran, V.M. (2015). Semantic enhanced social media sentiments for stock market prediction. *International Journal of Economics and Management Engineering* 9 (2): 678–682. https://doi.org/10.5281/zenodo.1100501.

23 Ji, X., Wang, J., and Yan, Z. (2021). A stock price prediction method based on deep learning technology. *International Journal of Crowd Science* 5 (1): 55–72. https://doi.org/10.1108/IJCS-05-2020-0012.

24 Puh, K. and Bagić Babac, M. (2023). Predicting stock market using natural language processing. *American Journal of Business* 38 (2): 41–61. https://doi.org/10.1108/AJB-08-2022-0124.

25 Li, X., Wu, P., and Wang, W. (2020). Incorporating stock prices and news sentiments for stock market prediction: a case of Hong Kong. *Information Processing & Management* 57 (5): 102212. https://doi.org/10.1016/j.ipm.2020.102212.

26 Moghar, A. and Hamiche, M. (2020). Stock market prediction using LSTM recurrent neural network. *Procedia Computer Science* 170: 1168–1173. https://doi.org/10.1016/j.procs.2020.03.049.

27 Patil, P.R., Parasar, D., and Charhate, S. (2021). A literature review on machine learning techniques and strategies applied to stock market price prediction. *Proceedings of the DDCIOT 2021*, Udaipur, India (20–21 March 2021), pp. 121–135.

28 Adhikari, R. and Agrawal, R.K. (2014). A combination of artificial neural network and random walk models for financial time series forecasting. *Neural Computing and Applications* 24: 1441–1449. https://doi.org/10.1007/s00521-013-1386-y.

29 Shahvaroughi Farahani, M. and Razavi Hajiagha, S.H. (2021). Forecasting stock price using integrated artificial neural network and metaheuristic algorithms compared to time series models. *Soft Computing* 25 (13): 8483–8513. https://doi.org/10.1007/s00500-021-05775-5.

30 Agarwal, D., Sheth, R., and Shekokar, N. (2021). Algorithmic trading using machine learning and neural network. In: *Computer Networks, Big Data and IoT: Proceedings of ICCBI 2020*, 407–421. Singapore: Springer.

31 Chhajer, P., Shah, M., and Kshirsagar, A. (2022). The applications of artificial neural networks, support vector machines, and long–short term memory for stock market prediction. *Decision Analytics Journal* 2: 100015. https://doi.org/10.1016/j.dajour.2021.100015.

32 Quah, T.S. and Srinivasan, B. (1999). Improving returns on stock investment through neural network selection. *Expert Systems with Applications* 17 (4): 295–301. https://doi.org/10.1016/S0957-4174(99)00041-X.

33 Wiranata, R.B. and Djunaidy, A. (2021). The stock exchange prediction using machine learning techniques: a comprehensive and systematic literature review. *Jurnal Ilmu Komputer dan Informatika* 14 (2): 91–112.

34 Mohammed, W.A. (2020). Challenges of stock prediction. In: *Valuation Challenges and Solutions in Contemporary Businesses* (ed. S.D. Köseoğlu), 234–252. IGI Global https://doi.org/10.4018/978-1-7998-1086-5.ch013.

35 Prasad, A., Bakhshi, P., and Guha, D. (2023). Forecasting the direction of daily changes in the India VIX index using deep learning. *IIMB Management Review* 35 (2): 149–163. https://doi.org/10.1016/j.iimb.2023.05.002.

36 Selvamuthu, D., Kumar, V., and Mishra, A. (2019). Indian stock market prediction using artificial neural networks on tick data. *Financial Innovation* 5 (1): 1–12. https://doi.org/10.1186/s40854-019-0131-7.

37 Qiu, M., Song, Y., and Akagi, F. (2016). Application of artificial neural network for the prediction of stock market returns: the case of the Japanese stock market. *Chaos, Solitons & Fractals* 85: 1–7. https://doi.org/10.1016/j.chaos.2016.01.004.

38 Liu, X., Guo, J., Wang, H., and Zhang, F. (2022). Prediction of stock market index based on ISSA-BP neural network. *Expert Systems with Applications* 204: 117604. https://doi.org/10.1016/j.eswa.2022.117604.

39 Hu, H., Tang, L., Zhang, S., and Wang, H. (2018). Predicting the direction of stock markets using optimized neural networks with Google Trends. *Neurocomputing* 285: 188–195. https://doi.org/10.1016/j.neucom.2018.01.038.

40 Dai, W., Wu, J.Y., and Lu, C.J. (2012). Combining nonlinear independent component analysis and neural network for the prediction of Asian stock market indexes. *Expert Systems with Applications* 39 (4): 4444–4452. https://doi.org/10.1016/j.eswa.2011.09.145.

41 Ding, G. and Qin, L. (2020). Study on the prediction of stock price based on the associated network model of LSTM. *International Journal of Machine Learning and Cybernetics* 11: 1307–1317. https://doi.org/10.1007/s13042-019-01041-1.

42 Rahman, M.O., Hossain, M.S., Junaid, T.S. et al. (2019). Predicting prices of stock market using gated recurrent units (GRUs) neural networks. *International Journal of Computer Science and Network Security* 19 (1): 213–222.

43 Li, Y. and Pan, Y. (2022). A novel ensemble deep learning model for stock prediction based on stock prices and news. *International Journal of Data Science and Analytics* 1–11. https://doi.org/10.1007/s41060-021-00279-9.

44 Eapen, J., Verma, A., and Bein, D. (2020). Improved big data stock index prediction using deep learning with CNN and GRU. *International Journal of Big Data Intelligence* 7 (4): 202–210. https://doi.org/10.1504/IJBDI.2020.113868.

45 Shahi, T.B., Shrestha, A., Neupane, A., and Guo, W. (2020). Stock price forecasting with deep learning: a comparative study. *Mathematics* 8 (9): 1441. https://doi.org/10.3390/math8091441.

46 Yu, P. and Yan, X. (2020). Stock price prediction based on deep neural networks. *Neural Computing and Applications* 32: 1609–1628. https://doi.org/10.1007/s00521-019-04212-x.

47 Jiang, W. (2021). Applications of deep learning in stock market prediction: recent progress. *Expert Systems with Applications* 184: 115537. https://doi.org/10.1016/j.eswa.2021.115537.

8

Predictive Analytics and Cybersecurity

Mohammed Sayeeduddin Habeeb

Department of Electrical Engineering, College of Engineering, King Khalid University, Abha, KSA

8.1 Introduction

In today's digital age, the increasing reliability on technology and interconnectedness has given rise to new threats and challenges in the realm of cybersecurity. Organizations and corporate enterprises worldwide are prime targets for cybercriminals and hackers, leaving business operations vulnerable to elevated levels of threats. In addition to deliberations regarding business strategies and the ever-changing market requirements, business proprietors are presently engaging in conversations revolving around cybersecurity frameworks and the degree of accomplishment achieved through their deployment. These frameworks play a pivotal role in managing cyber risks, pursuing dual objectives: the initial one involves devising strategies to counteract cyberattacks, while the second, of paramount importance, centers on devising strategies to proactively avert and substantially reduce the impact of these attacks.

Cyberattacks have become more sophisticated and pervasive, posing significant risks to individuals, businesses, and governments alike. Hackers continually modify their strategies, perpetually striving to maintain an edge. Consequently, it becomes imperative to establish a streamlined reporting mechanism and consistently apprise Chief Information Security Officers (CISOs) about emerging attack methodologies. This approach aims to minimize response times to the shortest possible duration [1].

As traditional security measures struggle to keep pace with rapidly evolving threats, the integration of predictive analytics with cybersecurity has emerged as a transformative approach to fortifying cybersecurity. Predictive analytics has the

Intelligent Techniques for Predictive Data Analytics, First Edition.
Edited by Neha Singh, Shilpi Birla, Mohd Dilshad Ansari, and Neeraj Kumar Shukla.

Published 2024 by John Wiley & Sons, Inc.

potential to identify a security breach even before it materializes. Similar to radar that reveals the adversary's impending arrival, advanced analytics indicate the potential occurrence and location of attacks. This empowers your business to issue alerts, fortify defenses, and prepare for the impending threat [2]. Predictive analytics equips you with the means to outsmart hackers, avoiding compromise after a battle is already lost. As intrusions become more numerous and intricate, your IT team is likely immersed in devising innovative and intelligent solutions. The technology community has long sought ways to safeguard sensitive data from malicious actors. However, emerging attacks now obfuscate their identifiers, rendering them immensely difficult to detect and combat. Predictive analytics possesses the potential to aid organizations in preemptively spotting security issues before they result in harm. By focusing on the "infection stage" rather than merely the "infection stage," businesses can predict future incidents and enhance preventative measures. Similar to IBM's mobile analyzer identifying flaws in mobile apps, hacker bots employ sophisticated analytics and extensive data to pinpoint vulnerabilities before launching an attack [3].

Through the utilization of predictive analytics and hacker bots, ongoing surveillance of operations takes place, offering crucial insights via autonomous learning analytics and detection methods. This empowers an organization to uncover threats even in the absence of precise assault signatures, addressing a current deficiency that proves inadequate against novel point-and-click vulnerabilities featuring distinct attack identifiers. Predictive analytics can instantaneously detect aberrations in data and traffic patterns, promptly notifying you of potential security issues prior to their occurrence. The application of predictive analytics spans diverse sectors, further elevating its prominence in aiding organizations to detect and address security susceptibilities.

8.2 Cybersecurity and Predictive Analysis

The realm of cybersecurity is in a perpetual state of evolution. Fresh threats surface, novel technologies are introduced, and security protocols undergo adjustments as organizations ascertain what proves effective and what does not. To maintain an edge over these dynamics, it becomes imperative to meticulously analyze data, discerning where security vulnerabilities reside and delineating actionable measures to abate them. Predictive analytics possesses the potential to be a transformative force within the realm of cybersecurity. Anticipatory insight constitutes a fundamental facet of security. It furnishes organizations with the ability to evaluate the probability of diverse scenarios materializing, informed by historical data [4]. This empowers organizations to forestall threats prior to their manifestation. Predictive analytics elevates predictive analysis to a higher echelon. Rather than confining itself to single prognostications, it engenders multiple hypotheses to ascertain the most probable outcome of a given event. This encompasses forecasting the likelihood of a specific eventuality based on an array of variables, encompassing historical data, user conduct, and external influences [2].

This chapter explores the intersection of predictive analytics and cybersecurity, presenting an in-depth analysis of its applications and benefits. Machine learning (ML) approaches are explained in brief and then proactive cybersecurity to real-time threat detection and cyber risk analytics, predictive analytics plays a crucial role in mitigating cyber threats and enhancing the overall security posture of organizations.

8.3 Machine Learning

ML classifier can used to learn datasets to classify algorithms and predict accurately. The goal of employing ML is to leverage prior cyberattacks to predict future attacks and make intelligent decisions. The ability to anticipate cyber issues prior to their occurrence will assist in mitigating the impact of detrimental activities. This distinct process contrasts with vulnerability scanning, which assesses metrics related to hostile actions. Given the absence of conspicuous indicators prior to a cyber event, predictive research must depend on unconventional signals that may or may not be linked to the potential target entity. The author in [5] introduced a tactics, techniques, and procedure (TTP) attack classifier by utilizing the vector space model that is frequently employed in the realm of information retrieval. This study presents an initial outcome that employs a Bayesian classifier to examine signals sourced from global events and social media [6].

On the contrary, authors employed an ML methodology to construct a classifier that automatically assigns labels of "Normal" or "Attack". There are two types of ML approaches based on the dataset supervised and unsupervised learning. In this section, the explanation of ML classification models for cybersecurity predictive analysis. Support vector machine (SVM), decision tree (DT), K-mean clustering, artificial neural network (ANN), fast learning network (FLN), K-nearest neighbor (KNN), and ensemble approaches are examples of ML algorithms used in cybersecurity. As shown in Figure 8.1.

8.3.1 Supervised and Unsupervised Learning

Two categories of ML exist: supervised and unsupervised. In supervised ML, one deals with a dataset containing input and output parameters. This approach accurately predicts system performance by training and testing with the provided

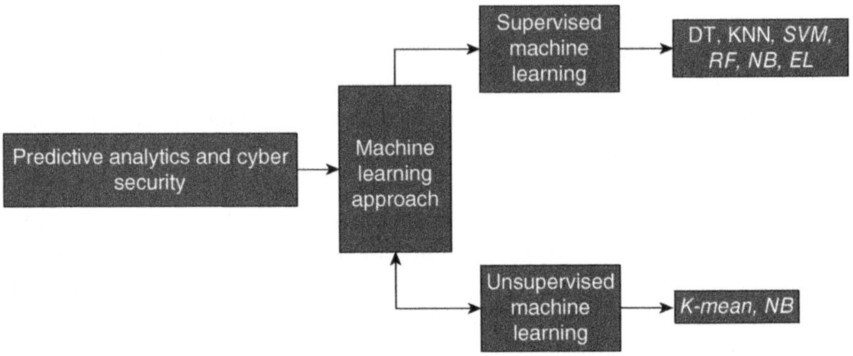

Figure 8.1 ML classification.

datasets. Unsupervised ML adopts a data-centric methodology aimed at identifying anomalies within data. It is utilized in scenarios where labeled data is unavailable, and the model autonomously discerns patterns based on inherent characteristics. Nonetheless, its precision is comparatively lower than that of the supervised approach, given that the system lacks predetermined datasets and is consequently employed for predicting unfamiliar outcomes [7].

Various methodologies exist for ML classification, employing algorithms, such as SVM, logistic regression (LG), DT, random forest (RF), Naïve Bayes, and CatBoost, all of which have demonstrated success within the realm of cybersecurity [8]. To discern the most suitable attributes for tree-based models, we utilized LG, SVM, and DT approaches. This enabled the identification of the optimal classification algorithm, enhancing the precision of predictive analytics for cyberattacks [9]. The classification of datasets may encounter a plethora of challenges. For instance, distinguishing attacks initiated through intelligent electronic devices in smart grid systems or classifying staff salaries based on qualifications, skills, and experience poses intricate classification difficulties [10].

8.3.2 ML Approach to Cybersecurity

In this section, we explore various ML approaches applicable to addressing cybersecurity challenges, along with their correlations to threats within predictive analysis. This data furnishes predictive cues enabling the anticipation of cyberattacks. ML approach for predictive analysis of data is shown in Figure 8.2. Data collected from the network nods are preprocessed and normalized in terms of normal and attacks category [11]. Some of the unwanted features are being discarded based on some of the feature selection algorithms, such as partial swam algorithm (PSA), whale optimization algorithms (WOAs), and many more, depends upon the requirements. After features are selected classifier model like SVM, DT, K-mean

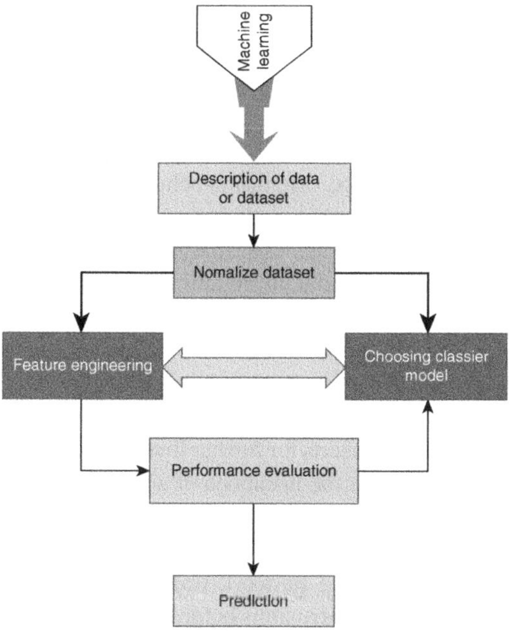

Figure 8.2 ML approach.

Table 8.1 Confusion matrix.

		Prediction class	
		Attack	**Normal**
Attack class	Attack	TP	TN
	Normal	FP	FN

TP, true positive; TN, true negative; FN, false negative; FP, false positive.

clustering, ANN, FLN, KNN are selected and then performance is calculated using confusion metrics [12] (Table 8.1).

Certainly, here are some examples of predictive data analysis in the context of cybersecurity using ML techniques:

- **Predictive Threat Intelligence:** ML models can analyze historical attack data, malware characteristics, and vulnerabilities to predict potential cyber threats. This enables organizations to proactively take measures to prevent attacks or minimize their impact [13].

- **User Authentication Behavior Anomalies**: By monitoring user behavior patterns, ML algorithms can identify anomalies that might indicate compromised accounts or insider threats. For instance, sudden access to sensitive data from an unusual location could trigger an alert [14].
- **Malware Detection:** ML algorithms can predict the likelihood of a file being malicious based on its attributes, such as file type, origin, and behavior. This helps in identifying and quarantining potential threats before they can cause harm [15].
- **Zero-Day Vulnerability Detection**: ML models can analyze software code and identify patterns that could indicate potential vulnerabilities. This predictive analysis can help developers patch or mitigate vulnerabilities before they are exploited [14].
- **Fraud Detection**: In financial systems, ML can predict fraudulent transactions by learning from historical fraud patterns and real-time data. Unusual transactions that deviate from normal behavior can trigger alerts for further investigation [16].
- **Phishing Attack Prevention**: ML can predict the likelihood of an email being a phishing attempt by analyzing its content, sender reputation, and context. This enables email filters to flag suspicious emails and reduce the risk of successful attacks [17].
- **Network Intrusion Detection**: By analyzing network traffic patterns and anomalies, ML can predict potential network intrusions before they occur. This can provide security teams with more time to respond and prevent breaches [17].
- **Advanced Persistent Threat (APT) Detection**: ML models can learn the behaviors of sophisticated APTs from historical data and predict their actions in real time. This allows for early detection and mitigation of ongoing attacks [18].
- **Patch Management Optimization**: ML can predict which systems are more vulnerable based on factors like system configuration and patch history. This helps IT teams prioritize patch deployment for maximum impact [19].
- **Data Leakage Prevention:** ML can predict data leakage by analyzing patterns of data movement within an organization. This can help identify potential data exfiltration attempts and prevent sensitive information from leaving the network [20].

Predictive data analysis with reference to ML empowers cybersecurity professionals with insights that allow them to anticipate and counter threats more effectively, ultimately enhancing an organization's overall security posture.

8.4 Proactive Cybersecurity and Real-Time Threat Detection

In the rapidly evolving landscape of cybersecurity, organizations are adopting proactive strategies that emphasize anticipation, prevention, and real-time response to cyber threats. Proactive cybersecurity focuses on staying ahead of attackers by predicting vulnerabilities, understanding attack patterns, and implementing measures that mitigate risks before they are exploited. Coupled with real-time threat detection, this approach creates a robust defense framework that helps organizations minimize damage and protect their digital assets.

8.4.1 Understanding Proactive Cybersecurity

Conventional cybersecurity measures primarily rely on reactive approaches, where security teams respond to incidents after they have occurred. However, this approach often leaves organizations vulnerable to new and emerging threats. Proactive cybersecurity takes a more preventative stance, employing predictive analytics to anticipate and counter potential vulnerabilities and threats before they materialize [21].

Proactive cybersecurity is rooted in the analysis of historical data, threat intelligence feeds, and past attack patterns. By understanding common attack vectors and emerging trends, predictive models can forecast potential future threats [22]. This empowers organizations to adopt preemptive measures, enhancing their security posture and effectively preventing cyberattackers from exploiting weaknesses.

- **Predictive Analysis:** Proactive cybersecurity employs predictive analytics to forecast potential threats and vulnerabilities. ML algorithms analyze historical data to identify patterns and anomalies that might indicate cyberattacks. By recognizing these patterns, organizations can predict and take preemptive measures against emerging threats [23].
- **Threat Intelligence:** Gathering and analyzing threat intelligence data provides insights into the TTPs of threat actors. This understanding helps organizations predict likely attack vectors, enabling them to deploy appropriate defenses and countermeasures [24].
- **Risk Assessments and Management:** A proactive approach involves assessing an organization's risk landscape, identifying critical assets, and assigning risk scores. This allows for the allocation of resources to protect high-value targets and address vulnerabilities proactively, minimizing potential impact [25, 26].
- **Vulnerability Management:** Regular vulnerability assessments and penetration testing help identify weaknesses before attackers exploit them. This involves scanning systems and applications for vulnerabilities and prioritizing patching based on risk [27].
- **Security Awareness Training:** Educating employees about cyber threats, social engineering, and best security practices empowers them to recognize and report suspicious activities. A well-informed workforce acts as an additional line of defense against attacks [28].

8.4.2 Real-Time Threat Detection

Real-time threat detection is a critical component of modern cybersecurity strategies, designed to identify and respond to cyber threats as they occur, minimizing potential damage and data breaches. This approach involves continuous monitoring of network activities, analyzing data in real time, and promptly responding to suspicious or malicious activities. By swiftly detecting and neutralizing threats, organizations can significantly reduce the impact of cyberattacks and protect sensitive information [29].

- **Intrusion Detection and Prevention System (IDPS)**: IDPS continuously monitors network traffic and system activities, looking for indicators of unauthorized access, malware, or other malicious activities. By instantly detecting anomalies, these systems trigger alerts or automated responses, preventing potential breaches [30].
- **Behavior Analysis:** Real-time behavioral analytics track user and entity behavior for deviations from normal patterns. Sudden changes can indicate compromised accounts or insider threats, prompting immediate investigation [31].
- **Security Information and Event Management (SIEM):** SIEM solutions aggregate and correlate security events from various sources in real time. By analyzing event data and applying rules, SIEM systems can detect and respond to threats as they occur [32].
- **Hunting Threat:** Proactive threat hunting involves actively seeking out potential threats and vulnerabilities within an organization's environment. This approach involves analyzing data, identifying anomalies, and investigating further to uncover potential risks [33].
- **Automated Response:** Real-time threat detection often involves automated responses such as isolating compromised systems, blocking malicious IP addresses, or quarantining suspicious files. These actions help contain threats and prevent them from spreading [34].

8.4.3 Use of Predictive Analytics for Threat Detection

Real-time threat detection is a cornerstone of modern cybersecurity. The ability to identify and respond promptly to cyber threats can significantly reduce the impact of attacks and prevent data breaches. Predictive analytics, fueled by advanced ML algorithms, plays a central role in enabling real-time threat detection [35, 36].

ML models continuously analyze vast amounts of data from various sources, such as network traffic, system logs, and user behavior. By learning from historical attack patterns, these models can detect anomalies and abnormal activities that may indicate potential cyber intrusions. Through constant adaptation and learning, predictive models remain agile, staying ahead of evolving threats and providing organizations with the tools to counter attacks in real time [37].

8.5 Network Security Analytics

Network security analytics is a comprehensive approach to cybersecurity that involves collecting, analyzing, and interpreting data from various network sources to detect and respond to potential cyber threats. This strategy goes beyond traditional security measures by utilizing advanced technologies and algorithms to identify anomalies, patterns, and behaviors indicative of malicious activities. By integrating diverse data sources and applying analytics, organizations can enhance their ability to identify and mitigate threats in real time [38].

8.5.1 Enhancing Network Security with Predictive Analytics

As the backbone of modern enterprises, networks serve as the lifeline connecting various devices and systems. However, they also represent a primary target for cyberattackers seeking to infiltrate organizations' critical assets and data. Predictive analytics plays a crucial role in enhancing network security by continuously analyzing network traffic and identifying potential threats.

As the backbone of modern enterprises, networks serve as the lifeline connecting various devices and systems. However, they also represent a primary target for cyberattackers seeking to infiltrate organizations' critical assets and data. Predictive analytics plays a crucial role in enhancing network security by continuously analyzing network traffic and identifying potential threats [39, 40].

8.5.2 The Role of ML in Network Security

ML plays a significant and evolving role in network security by providing advanced techniques to detect, prevent, and mitigate various types of cyber threats. ML algorithms have revolutionized network security analytics. Supervised learning models, trained on labeled data containing instances of normal and malicious network activity, can accurately classify and detect different types of cyber threats. Unsupervised learning techniques, on the contrary, are adept at identifying anomalies and previously unseen attack patterns, making them invaluable in combating zero-day threats [31].

Deep learning models, such as neural networks, excel at processing vast amounts of complex network data. Their ability to extract meaningful patterns and correlations from seemingly unrelated data points further strengthens network security analytics. Through continuous learning and adaptation, ML algorithms ensure that network security remains robust in the face of dynamic cyber threats [41, 42].

By monitoring network behavior, predictive models can establish baselines for normal activity. Any deviations from these baselines, such as sudden spikes in data transfer or unusual communication patterns, can be flagged as potential security breaches. This empowers network administrators to investigate and respond promptly to emerging threats, reducing the risk of data loss and service disruptions [43]. Here are some keyways in which ML contributes to network security:

Threat Detection and Anomaly Detection: ML algorithms can learn the normal patterns of network behavior and identify anomalies that deviate from these patterns. This can help detect various types of attacks, such as Distributed Denial of Service attacks, insider threats, and data exfiltration attempts. By continuously analyzing network traffic and user behavior, ML models can identify unusual activities and trigger alerts for further investigation.

Intrusion Detection and Prevention Systems (IDS/IPS): ML can enhance the capabilities of IDPS by enabling them to adapt and learn from new attack patterns. These systems can use ML to analyze network traffic in real time and detect known and unknown threats by identifying patterns that might not be easily discernible using traditional rule-based methods [44].

Malware Detection: ML can be employed to identify and classify malicious software based on their behavioral characteristics. By training on large datasets of known malware samples, ML models can learn to recognize patterns indicative of malware and help in detecting new and evolving threats [15].

Adaptive Security: ML allows security systems to adapt and evolve over time by continuously learning from new data. This adaptability is crucial for keeping up with rapidly evolving cyber threats and attack techniques [45].

However, it is important to note that while ML offers significant benefits, it is not a silver bullet and should be used in conjunction with other security measures. In addition, ML models can be vulnerable to adversarial attacks and biases, which require ongoing research and mitigation efforts.

8.6 Cyber Risk Analytics

Cyber risk analytics is the process of using data analysis techniques, such as statistical, predictive, and ML methods, to assess, quantify, and manage the potential risks associated with cyber threats and security incidents [46]. It involves collecting, processing, and interpreting various types of data to gain insights into an organization's cybersecurity posture, potential vulnerabilities, and the likelihood of different cyber risks occurring [1]. Here is a closer look at the key aspects of cyber risk analytics:

Data Collection: Cyber risk analytics relies on data from various sources, including network traffic, system logs, user behavior, threat intelligence feeds, historical incident data, and more. This data provides the foundation for analysis and modeling [3].

Risk Assessment: Cyber risk analytics involves assessing the severity and likelihood of different cyber risks. This assessment helps organizations understand the potential impact of various threats on their operations, assets, reputation, and compliance status [46].

Predictive Modeling: Predictive analytics uses historical data and ML algorithms to forecast future cyber threats and incidents. By identifying patterns and correlations in the data, predictive models can help organizations anticipate potential risks and take proactive measures [47].

Threat Detection: Cyber risk analytics plays a crucial role in detecting emerging threats and anomalies. By analyzing network behavior and system data, organizations can identify deviations from normal patterns, indicating potential security breaches or malicious activities [48].

8.6.1 Understanding Cyber Risk Management

Cyber risk analytics aims to assess an organization's exposure to cyber threats and quantify the potential financial losses associated with these risks. It involves a systematic approach to identifying and analyzing potential vulnerabilities, evaluating the likelihood of cyberattacks, and estimating their impact on the organization [49].

A risk-based approach in cybersecurity involves prioritizing resources and efforts based on the severity and probability of risks. By incorporating predictive analytics into cyber risk management, organizations can make informed decisions and allocate resources effectively to address the most significant threats [50].

8.6.2 Predictive Modeling for Cyber Risk Assessment

Predictive modeling lies at the heart of cyber risk analytics, providing organizations with the tools to predict and prepare for cyber threats effectively. By analyzing historical data and threat intelligence, predictive models estimate the likelihood of various cyberattack scenarios and the potential financial impact on the organization [51].

Monte Carlo simulations are widely used in cyber risk assessment. By simulating thousands of potential cyberattack scenarios, organizations can understand the range of potential losses they may face. Bayesian networks offer a more comprehensive view of cyber risks by considering the interdependencies between different risk factors, providing a more accurate risk assessment [26].

8.7 Impact of Predictive Analytics on the Cybersecurity Landscape

Utilizing security analytics, organizations can counter cyberattacks by pinpointing indicators linked to malicious conduct. Analysts can enhance their readiness for such scenarios by employing predictive insights, granting them additional time to respond and strategize. For instance, imagine an organization notices a surge in user logins originating from a specific country. In response, they could construct a predictive model to gauge whether this pattern signifies potential malicious intent [52].

Predictive analytics also empower security teams to anticipate the timing of attacks and make informed preparations. This proactive approach may involve raising awareness about emerging threats, equipping team members to swiftly detect and address potential issues. Furthermore, predictive analytics can aid in foreseeing the success of an imminent attack, prompting immediate responses and countermeasures [2].

Consider a scenario where an organization detects a heightened frequency of user logins from a particular country, coupled with signs of malicious behavior. In this context, crafting a predictive model to forecast potential malevolent actions allows security teams to brace themselves and potentially thwart a harmful incident before it transpires [4].

8.8 Challenges in Applying Predictive Analytics to Cybersecurity

The utilization of predictive analytics in the realm of cybersecurity presents certain challenges that necessitate careful consideration. While predictive analytics holds substantial potential to enhance security strategies, several inherent hurdles must be addressed to ensure its effective implementation and meaningful results.

Data Accuracy and Quality: The accuracy and quality of the data used to train predictive models play a pivotal role in their effectiveness. If the input data is incomplete, outdated, or erroneous, the resulting predictions may be unreliable or misleading. Ensuring the integrity of data sources is paramount to deriving accurate insights [2].

Bias and Irrelevant Variable: Predictive models are susceptible to bias if they incorporate irrelevant variables or if the training data exhibits inherent biases. Such biases can lead to skewed predictions and potentially cause misguided decisions. Thorough data preprocessing and feature selection are vital to mitigate this challenge [4].

Model Over Fitting and Generalization: Striking the right balance between models that accurately fits the training data (over fitting) and one that can generalize to new data (under fitting) is complex. Over fitting can lead to models that perform well on training data but fail to generalize to new, unseen data [38].

Feature Selection and Relevance: Determining which features or variables to include in the model requires careful consideration. Including irrelevant or redundant variables can lead to noisy models that provide inaccurate predictions [53].

Dynamic and Improving Threats: The rapidly evolving nature of cyber threats demands models that can adapt and remain effective over time. Predictive models must be regularly updated and retrained to account for emerging attack techniques and tactics [54].

Data Privacy: The use of predictive analytics involves handling potentially sensitive data. Ensuring compliance with data protection regulations and maintaining the privacy of individuals is essential and may introduce complexities in data collection and usage [55].

Lack of Understanding: Predictive analytics may identify patterns or anomalies, but understanding the context behind these findings is essential. False positives can occur if patterns are misinterpreted as threats due to a lack of context [56].

Human Factors: Effective utilization of predictive analytics requires a skilled workforce capable of interpreting and acting upon the insights provided. Insufficient training and understanding of these tools can limit their effectiveness [37, 57].

False Positive and Negatives: Predictive models may produce false positives (indicating a threat that is not present) or false negatives (failing to detect an actual threat), which can lead to inefficiencies in security operations [53].

Addressing these challenges requires a holistic approach that involves data governance, model validation, ongoing monitoring, and a thorough understanding of both predictive analytics and the cybersecurity domain. While predictive analytics holds significant potential for enhancing cybersecurity, organizations must navigate these obstacles to fully harness its benefits.

8.9 Conclusion

The integration of predictive analytics into cybersecurity represents a paradigm shift in how organizations approach and combat cyber threats. By adopting a proactive stance, organizations can anticipate and prevent potential attacks before they materialize. Real-time threat detection powered by predictive analytics enables swift responses to emerging threats, minimizing the impact of cyberattacks.

Incorporating predictive analytics into network security empowers organizations to identify anomalies and potential breaches promptly, safeguarding their digital infrastructure and critical assets. Moreover, cyber risk analytics helps organizations prioritize their efforts and resources, enabling them to address the most significant threats first.

As cyber threats continue to evolve, the role of predictive analytics in cybersecurity will only become more crucial. By harnessing the power of data-driven insights and advanced ML techniques, organizations can stay one step ahead of cyber adversaries and build a more resilient and secure digital future. The fusion of predictive analytics and cybersecurity is the key to achieving a safer and more secure digital ecosystem for all.

References

1 Srivastava, N. and Jaiswal, U.C. (2019). Big data analytics technique in cyber security: a review. *Proceedings of the 3rd International Conference on Computing Methodologies and Communication, ICCMC 2019*, pp. 579–585, Erode, India (27–29 March 2019). New York: IEEE. https://doi.org/10.1109/ICCMC.2019.8819634.

2 Bhardwaj, A. and Kaushik, K. (2022). Predictive analytics-based cybersecurity framework for cloud infrastructure. *International Journal of Cloud Applications and Computing* 12 (1): 1–20. https://doi.org/10.4018/IJCAC.297106.

3 Yeboah-Ofori, A., Islam, S., Lee, S.W. et al. (2021). Cyber threat predictive analytics for improving cyber supply chain security. *IEEE Access* 9: 94318–94337. https://doi.org/10.1109/ACCESS.2021.3087109.

4 Eastman, R., Versace, M. and Webber, A. (2015). Big data and predictive analytics: on the cybersecurity front line IDC white paper.

5 Gallagher, B. and Eliassi-Rad, T. (2008). Classification of HTTP attacks: a study on the ECML/PKDD 2007 discovery challenge. *Center for Advanced Signal and Image Sciences (CASIS) Workshop*. http://www.briangallagher.net/pubs/gallagher-llnltr-2009.pdf (accessed 2 August 2023).

6 Okutan, A., Yang, S.J., and McConky, K. (2017). Predicting cyber attacks with Bayesian networks using unconventional signals. *Proceedings of the 12th Annual Conference on Cyber and Information Security Research (CISRC '17). Association for Computing Machinery*, New York, NY, USA, Article 13. https://doi.org/10.1145/3064814.3064823.

7 Ahmad, Z., Khan, A.S., Shiang, C.W. et al. (2021). Network intrusion detection system: a systematic study of machine learning and deep learning approaches. *Transactions on Emerging Telecommunications Technologies* 32 (1): https://doi.org/10.1002/ETT.4150.

8 Giancarlo, E. and Villano, V. (2018). *Classification of Logs Using Machine Learning Technique*. NTNU.

9 Husák, M. and Kašpar, J. (2018). Towards predicting cyber attacks using information exchange and data mining. *2018 14th International Wireless Communications and Mobile Computing Conference, IWCMC 2018*, pp. 536–541, Limassol, Cyprus (25–29 June 2018). New York: IEEE. https://doi.org/10.1109/IWCMC.2018.8450512.

10 Alberto, B. and Massaron, L. (2016). *Python Data Science Essentials: Become an Efficient Data Science Practitioner by Understanding Python's Key Concepts*. Packt Publishing.

11 Habeeb, M.S. and Babu, T.R. (2022). Network intrusion detection system: a survey on artificial intelligence-based techniques. *Expert Systems* 39 (9): e13066. https://doi.org/10.1111/EXSY.13066.

12 Deng, X., Liu, Q., Deng, Y., and Mahadevan, S. (2016). An improved method to construct basic probability assignment based on the confusion matrix for classification problem. *Information Sciences* 340–341: 250–261. https://doi.org/10.1016/j.ins.2016.01.033.

13 Sree, V.S., Koganti, C.S., Kalyana, S.K., and Anudeep, P. (2021). Artificial intelligence based predictive threat hunting in the field of cyber security. *2021 2nd Global Conference for Advancement in Technology, GCAT 2021*, Bangalore, India (1–3 October 2021). New York: IEEE. https://doi.org/10.1109/GCAT52182.2021.9587507.

14 Pannell, G. and Ashman, H. (2010). User modelling for exclusion and anomaly detection: a behavioural intrusion detection system. In: De Bra, P., Kobsa, A., Chin, D. (eds) User Modeling, Adaptation, and Personalization. UMAP 2010. Lecture Notes in Computer Science, vol 6075. Springer, Berlin, Heidelberg.

15 Yeboah-Ofori, A. and Boachie, C. (2019). Malware attack predictive analytics in a cyber supply chain context using machine learning. *Proceedings – 2019 International Conference on Cyber Security and Internet of Things, ICSIoT 2019* (66–73), Accra, Ghana (29–31 May 2019). New York: IEEE. https://doi.org/10.1109/ICSIOT47925.2019.00019.

16 Singla, A. and Jangir, H. (2020). 'A comparative approach to predictive analytics with machine learning for fraud detection of realtime financial data. *Proceedings - 2020 International Conference on Emerging Trends in Communication, Control and Computing, ICONC3 2020*, Lakshmangarh, India (21–22 February 2020). New York: IEEE. https://doi.org/10.1109/ICONC345789.2020.9117435.

17 Raja, S.E. and Ravi, R. (2020). A performance analysis of software defined network based prevention on phishing attack in cyberspace using a deep machine learning with CANTINA approach (DMLCA). *Computer Communications* 153: 375–381. https://doi.org/10.1016/J.COMCOM.2019.11.047.

18 Chen, P., Desmet, L., and Huygens, C. (2014). A study on advanced persistent threats. In: *Communications and Multimedia Security. CMS 2014. Lecture Notes in Computer Science*, vol. 8735 (ed. B. De Decker and A. Zúquete). Berlin, Heidelberg: Springer https://doi.org/10.1007/978-3-662-44885-4_5/COVER.

19 Grotto, A. and Dempsey, J. (2021). Vulnerability disclosure and management for AI/ML systems: a working paper with policy recommendations. *SSRN Electronic Journal.* https://doi.org/10.2139/SSRN.3964084.

20 Balajee, J.M. (2018). Data wrangling and data leakage in machine learning for healthcare. https://papers.ssrn.com/abstract=3708142 (accessed 15 August 2023).

21 Craig, A.N., Shackelford, S.J., and Hiller, J.S. (2015). Proactive cybersecurity: a comparative industry and regulatory analysis. *American Business Law Journal* 52 (4): 721–787. https://doi.org/10.1111/ABLJ.12055.

22 Offner, K.L., Sitnikova, E., Joiner, K., and MacIntyre, C.R. (2021). Towards understanding cybersecurity capability in Australian healthcare organisations: a systematic review of recent trends, threats and mitigation. *Health Security Intelligence* 92–121. https://doi.org/10.4324/9781003245483-8.

23 Colbaugh, R. and Glass, K. (2011). Proactive defense for evolving cyber threats. *Proceedings of 2011 IEEE International Conference on Intelligence and Security Informatics, ISI 2011*, pp. 125–130, Beijing, China (10–12 July 2011). New York: IEEE. https://doi.org/10.1109/ISI.2011.5984062.

24 Sun, N., Ding, M., Jiang, J. et al. (2023). Cyber threat intelligence mining for proactive cybersecurity defense: a survey and new perspectives. *IEEE Communications Surveys and Tutorials.* https://doi.org/10.1109/COMST.2023.3273282.

25 Gunes, B., Kayisoglu, G., and Bolat, P. (2021). Cyber security risk assessment for seaports: a case study of a container port. *Computers & Security* 103: 102196. https://doi.org/10.1016/J.COSE.2021.102196.

26 Ferreira, D.J., Mateus-Coelho, N., and Mamede, H.S. (2023). Methodology for predictive cyber security risk assessment (PCSRA). *Procedia Computer Science* 219: 1555–1563. https://doi.org/10.1016/J.PROCS.2023.01.447.

27 Tang, M.J., Alazab, M., and Luo, Y. (2019). Big data for cybersecurity: vulnerability disclosure trends and dependencies. *IEEE Transactions on Big Data* 5 (3): 635–647. https://doi.org/10.1109/TBDATA.2017.2723570.

28 Kemper, G. (2021). Improving employees' cyber security awareness. *Computer Fraud & Security* 2019 (8): 11–14. https://doi.org/10.1016/S1361-3723(19)30085-5.

29 Cichonski, P., Millar, T., Grance, T., and Scarfone, K. (2012) 'Computer Security Incident Handling Guide Recommendations of the National Institute of Standards and Technology'. National Institute of Standards and Technology. https://doi.org/10.6028/NIST.SP.800-61r2.

30 Borkar, A., Donode, A. and Kumari, A. (2018). A survey on intrusion detection system (IDS) and internal intrusion detection and protection system (IIDPS). *Proceedings of the International Conference on Inventive Computing and Informatics, ICICI 2017*, pp. 949–953, Coimbatore, India (23–24 November 2017). New York: IEEE. https://doi.org/10.1109/ICICI.2017.8365277.

31 Soni, J., Prabakar, N. and Upadhyay, H. (2019). Behavioral analysis of system call sequences using LSTM seq-seq, cosine similarity and jaccard similarity for real-time anomaly detection. *Proceedings - 6th Annual Conference on Computational Science and Computational Intelligence, CSCI 2019*, pp. 214–219, Las Vegas, NV (5–7 December 2019). New York: IEEE. https://doi.org/10.1109/CSCI49370.2019.00043.

32 González-Granadillo, G., González-Zarzosa, S., and Diaz, R. (2021). Security information and event management (SIEM): analysis, trends, and usage in critical infrastructures. *Sensors* 21 (14): 4759. https://doi.org/10.3390/S21144759.

33 Mavroeidis, V. and Jøsang, A. (2018). Data-driven threat hunting using sysmon. *2nd International Conference on Cryptography, Security and Privacy (ICCSP 2018). Association for Computing Machinery*, New York, NY, USA, pp. 82–88. https://doi.org/10.1145/3199478.3199490.

34 Kanoun, W., Cuppens-Boulahia, N., Cuppens, F., and Araujo, J. (2008). Automated reaction based on risk analysis and attackers skills in intrusion detection systems. *Proceedings 2008 3rd International Conference on Risks and Security of Internet and Systems, CRiSIS 2008*, pp. 117–124, Tozeur, Tunisia (28–30 October 2008). New York: IEEE. https://doi.org/10.1109/CRISIS.2008.4757471.

35 More, R., Unakal, A., Kulkarni, V., and Goudar, R.H. (2017). Real time threat detection system in cloud using big data analytics. *RTEICT 2017 – 2nd IEEE International Conference on Recent Trends in Electronics, Information and Communication Technology, Proceedings*, Bangalore, India (19–20 May 2017), pp. 1262–1264. https://doi.org/10.1109/RTEICT.2017.8256801.

36 Masood, A. and Hashmi, A. (2019). AIOps: Predictive Analytics & Machine Learning in Operations. In: *Cognitive Computing Recipes* (ed. A. Masood and A. Hashmi), 359–382. Apress https://doi.org/10.1007/978-1-4842-4106-6_7.

37 Mayhew, M., Atighetchi, M., Adler, A., and Greenstadt, R. (2015). Use of machine learning in big data analytics for insider threat detection. *Proceedings – IEEE Military Communications Conference MILCOM*, 2015-December, pp. 915–922, Tampa, FL (26–28 October 2015). New York: IEEE. https://doi.org/10.1109/MILCOM.2015.7357562.

38 Mahmood, T. and Afzal, U. (2013). Security analytics: big data analytics for cybersecurity: a review of trends, techniques and tools. *Conference Proceedings - 2013 2nd National Conference on Information Assurance, NCIA 2013*, pp. 129–134 Rawalpindi, Pakistan (11–12 December 2013). New York: IEEE. https://doi.org/10.1109/NCIA.2013.6725337.

39 Sengan, S., Subramaniyaswamy, V., Nair, S.K. et al. (2020). Enhancing cyber–physical systems with hybrid smart city cyber security architecture for secure public data-smart network. *Future Generation Computer Systems* 112: 724–737. https://doi.org/10.1016/J.FUTURE.2020.06.028.

40 Priyadarshini, I., Kumar, R., Tuấn, L.M. et al. (2021). A new enhanced cyber security framework for medical cyber physical systems. *Software-Intensive Cyber-Physical Systems* 35 (3–4): 159–183. https://doi.org/10.1007/S00450-021-00427-3/METRICS.

41 Tavallaee, M., Bagheri, E., Lu, W., and Ghorbani, A.A. (2009). A detailed analysis of the KDD CUP 99 data set. *IEEE Symposium on Computational Intelligence for Security and Defense Applications, CISDA 2009*, Ottawa, ON (8–10 July 2009). New York: IEEE. https://doi.org/10.1109/CISDA.2009.5356528.

42 Deng, J., Zhang, Z.; Marchi, E., and Schuller, B. (2013). Sparse autoencoder-based feature transfer learning for speech emotion recognition. *Proceedings – 2013 Humaine Association Conference on Affective Computing and Intelligent Interaction, ACII 2013*, pp. 511–516, Geneva, Switzerland (2–5 September 2013). New York: IEEE. https://doi.org/10.1109/ACII.2013.90.

43 Moustafa, N. and Slay, J. (2015). UNSW-NB15: A comprehensive data set for network intrusion detection systems (UNSW-NB15 network data set). *2015 Military Communications and Information Systems Conference, MilCIS 2015 – Proceedings*, Canberra, ACT, Australia (10–12 November 2015). New York: IEEE https://doi.org/10.1109/MilCIS.2015.7348942.

44 Akbar, S., Srinivasa Rao, T., and Ali Hussain, M. (2016). A hybrid scheme based on big data analytics using intrusion detection system. *Indian Journal of Science and Technology* 9 (33): 1–4. https://doi.org/10.17485/ijst/2016/v9i33/97037.

45 Lin, H., Yan, Z., and Fu, Y. (2019). Adaptive security-related data collection with context awareness. *Journal of Network and Computer Applications* 126: 88–103. https://doi.org/10.1016/J.JNCA.2018.11.002.

46 Radanliev, P., De Roure, D., Page, K. et al. (2020). Design of a dynamic and self-adapting system, supported with artificial intelligence, machine learning and real-time intelligence for predictive cyber risk analytics in extreme environments – cyber risk in the colonisation of Mars. *Safety in Extreme Environments* 2 (3): 219–230. https://doi.org/10.1007/S42797-021-00025-1/FIGURES/1.

47 Radanliev, P., De Roure, D., Walton, R. et al. (2020). Artificial intelligence and machine learning in dynamic cyber risk analytics at the edge. *SN Applied Sciences* 2 (11): 1–8. https://doi.org/10.1007/S42452-020-03559-4/FIGURES/2.

48 Gheyas, I.A. and Abdallah, A.E. (2016). Detection and prediction of insider threats to cyber security: a systematic literature review and meta-analysis. *Big Data Analytics* 1 (1): 1–29. https://doi.org/10.1186/S41044-016-0006-0.

49 Refsdal, A., Solhaug, B., and Stølen, K. (2015). *Cyber-risk Management*', SpringerBriefs in Computer Science, 33–47. https://doi.org/10.1007/978-3-319-23570-7_5/COVER.

50 Lee, I. (2020). Internet of Things (IoT) cybersecurity: literature review and IoT cyber risk management. *Future Internet* 12 (9): 157. https://doi.org/10.3390/FI12090157.

51 Tubío Figueira, P., López Bravo, C., and Rivas López, J.L. (2020). Improving information security risk analysis by including threat-occurrence predictive models. *Computers & Security* 88: 101609. https://doi.org/10.1016/J.COSE.2019.101609.

52 Sharma, P. and Dash, B. (2023). Impact of big data analytics and ChatGPT on cybersecurity. *2023 4th International Conference on Computing and Communication Systems, I3CS 2023*, Shillong, India (16–18 March 2023). New York: IEEE. https://doi.org/10.1109/I3CS58314.2023.10127411.

53 Soni, V.D. (2020). Challenges and solution for artificial intelligence in cybersecurity of the USA. *SSRN Electronic Journal* https://doi.org/10.2139/SSRN.3624487.

54 Sarker, I.H. (2022). Machine learning for intelligent data analysis and automation in cybersecurity: current and future prospects. In: *Annals of Data Science* (ed. Y. Shi), 1–26. https://doi.org/10.1007/S40745-022-00444-2/FIGURES/6.

55 Keim, D.A., Mansmann, F., Schneidewind, J., and Ziegler, H. (2006). Challenges in visual data analysis. *Proceedings of the International Conference on Information Visualisation*, pp. 9–14, London, UK (5–7 July 2006). New York: IEEE. https://doi.org/10.1109/IV.2006.31.

56 Moustafa, N. (2021). A systemic IoT–Fog–Cloud architecture for big data analytics and cyber security systems: a review of fog computing. *Secure Edge Computing* 41–50. https://doi.org/10.1201/9781003028635-4.

57 Amalina, F., Hashem, I.A.T., Azizul, Z.H. et al. (2020). Blending big data analytics: review on challenges and a recent study. *IEEE Access* 8: 3629–3645. https://doi.org/10.1109/ACCESS.2019.2923270.

9

Precision Agriculture and Predictive Analytics

Enhancing Agricultural Efficiency and Yield

Nafees Akhter Farooqui[1], *Mohd. Haleem*[2], *Wasim Khan*[1], *and Mohammad Ishrat*[1]

[1] *Department of CSE, Koneru Lakshmaiah Education Foundation, Vaddeswaram, Andhra Pradesh, India*
[2] *Department of Computer Science, Era University, Lucknow, Uttar Pradesh, India*

9.1 Introduction

The term "agriculture system" refers to the complex interplay of various components, processes, and activities involved in producing, managing, and distributing agricultural products. An agriculture system, often called an agricultural system or farming system, encompasses the intricate network of processes, practices, and technologies involved in producing food, fiber, and other agricultural products [1]. It is a comprehensive framework that integrates various elements to ensure efficient and sustainable production while considering economic, environmental, and social factors. Agriculture systems have evolved over time to adapt to changing conditions and technological advancements. Precisely due to the many difficulties farmers currently confront, the technological evolution of the agriculture sector has become a critical issue [2]. The use of remote monitoring and control technologies to improve agricultural efficiency and competitiveness is expanding quickly. This may be attainable with the use of smart farming and precision agriculture. These two phrases describe the process of achieving fine-grid crop management using cutting-edge technologies in conventional farming methods. To increase competitiveness and profitability, smart farming technologies can offer farmers valuable environmental data from the cultivated areas. These kinds of technological advancements can improve nearly every facet of the agriculture system. Unmanned aerial vehicles (UAVs) and other forms of

Intelligent Techniques for Predictive Data Analytics, First Edition.
Edited by Neha Singh, Shilpi Birla, Mohd Dilshad Ansari, and Neeraj Kumar Shukla.

Published 2024 by John Wiley & Sons, Inc.

intelligent sensors are at the forefront of remote sensing technology, which will shape the future of precision agriculture [3]. An external stimulation, or input signal, is transformed by a sensor into a measured output signal. Macroscopic quantities can be converted to electrically quantifiable quantities using sensor devices. When an electrical signal is processed and transformed into a uniform signal with predefined parameters. Changes in the sensor's environment or the use of a specific analyzer can cause observable changes in the sensor's characteristics [4]. UAVs are airplanes that fly without a human pilot. Instead, they can fly independently or be controlled from a distance [5]. Drones for remote monitoring are part of unmanned aerial systems (UASs). It comprises all the hardware and software required to fly and manage UAVs and collect data. Earth observation (EO) is aided by UAVs in three ways: precision is increased, field monitoring is increased and expanded, and significant areas that are difficult to access are covered. The Internet of Things (IoT) paradigm includes devices like sensors and UAVs. The IoT is a cutting-edge communications network that uses a plethora of dispersed smart devices connected by a single wide area network [6]. Its defining feature is smart devices' capacity to detect and alert users to unfold events in real time. The computing abilities, resource availability, and capacity for independent communication of smart objects and devices are minimal. However, the widespread implementation of intelligent farming is hindered by the absence of models that would instruct farmers on how to assemble their own IoT-based systems for monitoring [6]. The seamless integration of the needed components is greatly hampered by the often dense and heterogeneous nature of IoT deployments. However, the adoption and deployment of smart devices into our daily activities can be facilitated by an effectively built architecture based on recently upgraded IoT technology.

Precision farming can be thought of as a self-running process that gathers data and sends it to systems that analyze it. The use of data mining in agriculture has become very important because it can look through huge amounts of data to find new information and improve the way things are done now. In this scenario, the soil profile is one of the things you need to know before you can make good choices about agriculture. Sampling the soil is a good way to get this helpful knowledge. But it is expensive and takes a lot of time. Also, it is not always necessary to test the dirt on every field because the conditions can be the same as on fields locally. Precision agriculture, also known as site-specific farming or satellite farming, is a modern method of agricultural management that utilizes technology to optimize crop production [7]. This chapter explores the integration of predictive analytics into precision agriculture and its potential to enhance agricultural efficiency and yield. By utilizing data-driven insights, farmers can make informed decisions about irrigation, fertilization, pest control, and other critical factors that influence crop health and productivity [8].

Precision agriculture and its components involve the use of advanced technologies such as global positioning systems (GPS), geographic information systems (GIS), remote sensing, and data analytics [9]. These tools enable farmers to collect and analyze data. By understanding the spatial and temporal variations within their fields, farmers can tailor their management practices accordingly, leading to improved resource allocation and increased productivity.

9.2 Background

Significant advancement has been made in the field of precision agriculture through the utilization of many techniques, such as the IoT and machine learning (ML) for decision support systems and data analytics, among others. There are three subsections of background explanations: (i) IoT and networks, (ii) predictive analytics, and (iii) applications of Precision agriculture systems. The composition of precision in different fields is systematically stated as follows:

9.2.1 IoT and Networks

"Wireless sensor networks (WSN)" find application in several agronomic contexts, enabling remote monitoring of ambient and soil conditions to facilitate the prediction of crop well-being. These networks are used to forecast watering schedules for agricultural fields based on humidity, temperature, soil moisture, and salinity. Researchers have developed scalable network architectures, IoT-based control systems, and IoT frameworks for various applications. Ahmed et al. [10] proposed a scalable network architecture for monitoring and controlling agriculture fields. The system combines long-distance networks and fog computing solutions for high throughput.

Muangprathub et al. [11] applied IoT improves crop yields, quality, and costs in agriculture. They developed a system for optimal crop development called Smartnode, which uses hardware and software to monitor agro-climatic parameters. Oliver et al. [12] proposed the SEnviro platform to monitor vineyard fields using edge computing. Elkarim et al. [13] suggested a model using IoT and ML to predict crop diseases using environmental sensors. Jawad et al. [14] reviewed agriculture applications based on WSNs and compared several wireless technologies and protocols.

Sikeridis et al. [15] incorporate ML with IoT to improve energy efficiency in public safety networks. Hsu et al. [16] proposed an innovative enhancement of public-to-physical connectivity and IoT processing speeds via an on-demand approach based on cloud-based IT platforms. To integrate and automate cultivating using limited informational resources, cloud integration can be employed for large-area data collecting and analytics. In precision agriculture, IoT systems have been developed using the "Nitrogen-Phosphorus-Potassium (NPK)" sensor, which uses colorimetric principles to monitor and analyze soil nutrients. Google's cloud databases provide convenient access to data whenever it is required. Nutrients are determined using a fuzzy logic method applied to the gathered data. The developed IoT system is most useful for agrarians who have high crop yields. Numerous researchers have explored the potential of utilizing the IoT in "smart" farms, including remote greenhouse monitoring, efficient crop monitoring, and data collection from agricultural fields [17]. Some authors have used various technologies, such as "WSN," "IoT," "Weather Stations," and "Drone-enabled Camera," to collect and analyze data from agricultural fields. The platform SmartFarmNet was developed to process data related to various parameters and forecast crop status [18].

9.2.2 Predictive Analytics

Earlier information technology, conventional techniques for detecting crop diseases and pests were cumbersome and often resulted in human error. ML, which uses data analytics to learn from experiences, has become increasingly useful in forecasting crop diseases and pests. Techniques used include "Artificial Neural Networks (ANN)," "Support Vector Machines (SVM)," and "Fuzzy Technology" for recognition [19].

Several studies used ML methods to predict crop diseases and pests, such as classifying apple diseases using algorithms like random forest, K nearest neighbor, SVM, and naïve bayes. Other studies used IoT and ML to improve the service of limited assets, for instance, Ramesh et al. [20] used IoT and ML techniques to predict late blight disease in potatoes, using environmental parameters to detect the level and risk of blight. Farooqui et al. [21] used image processing and ML algorithms to predict site-specific yield with an accuracy of over 80%.

Cetişli and Büyükçingir [22] suggested a novel apple scab early detection forecasting technique employing AI and time series forecasting. Balducci et al. [23] emphasized the importance of using continuous data for decision-making. Hamad et al. [24] highlighted the benefits of an intelligent system for acquiring practical agriculture facts such as soil moisture, humidity, and temperature.

The importance of agriculture in feeding the world's population is immeasurable. Even though humanity continues to grow, agricultural land remains unchanged [25]. It is critical to identify the chemistry between category, soil qualities, fertilizer, and climate when designing and maintaining a farm. The diverse character of these elements makes it impossible to select them effectively with the conventional method. Kan et al. [25] claimed that variations in these sources had a deleterious effect on agricultural production [11]. Global warming has been shown to have a significant negative impact on agricultural productivity. Low-agricultural output remains a problem for farmers since farming resources are inconsistent and poorly managed. Due to repeated use and a lack of understanding about the necessary elements to complete the wasted nutrients, the land used for cultivation continuously degrades, making decreased production even more worrying. As a result, the increasing number of mouths to feed necessitates new strategies. There is a critical need in agriculture for a method of accurately predicting soil chemical characteristics, fertilizer, and climate conditions that will allow crops to thrive. In addition, high crop production relies heavily on anticipating the optimal combination of these resources [26]. Crop yield has been significantly affected, particularly in developing nations, due to a lack of information about the applicability of these factors to increase agricultural output. The type of soil and manure that is suitable for crops needs to be determined in an efficient and automated manner. These data are urgently needed in agriculture to prevent a catastrophic food shortage [27] due to a shortage of yield production.

In addition, precision agriculture occurred as a breakthrough in the farming industry, with the goal of automating farming procedures to increase yield quantity and quality for long-term food security. To achieve the greatest yield and quality, "smart farms" implement technological solutions and AI-based management systems [28]. The development of precision agriculture has brought innovation to the agricultural sector, automating farming processes to achieve higher yields of both quality and quantity for natural diet production. Precision farming involves technologies to enhance farm management and improve crop production [29]. The initial agricultural revolution was agricultural mechanization (the use of tools and machinery), and the second was the green revolution, which focused on the enhancement of seeds for crops and animal breeds [30]. Precision agriculture and smart farming share certain similarities, but both are not the same. Both can help close the knowledge gap and are primarily concerned with increasing harvest production. John Deere is widely credited with conceptualizing the concept of digital agriculture in 1990 when sophisticated instruments enabled planting and fertilizer application [30]. This new approach was developed to increase crop yields and encourage more people to take up farming. For accurate crop yield predictions, smart farmers collect and analyze data on different parameters such as soil quality, chemical requirements, and weather. This information is used to develop automated oversight and administration models for agricultural resources [23]. Particularly essential are innovative agricultural practices and/or precision farming in developing countries like Nigeria, where agriculture is largely done in its traditional forms.

ML methods are ideal for reliably forecasting these crucial agricultural aspects. In this regard, Ali et al. [31] emphasized ML in protective smart farming prediction. Pandiarajaa [32] said that the automated benefits of ML might be applied to every area of human activity. As the largest single sector of the global economy and a vital food source, agriculture is also in need of these advancements. Smart farming, also known as precision agriculture, relies heavily on ML, which is at the heart of the agricultural digitization movement [31]. ML strategies have been proposed to address the varying conditions of crops, soils, fertilizers, and weather [33]. Solutions for smart farming prediction have also been proposed using deep learning and ensemble models [34]. These ML algorithms improve yields and quality, expanding agriculture's role in the global economy. When put into practice, learning models that make use of data from the real world to train can function independently of human oversight [33]. The authors [4] presented an IoT-based system for farm management via web-based apps that are reachable via internet-enabled mobile devices. Harvesting calculation, identifying potential cultivation concerns, and recommending enhancements are all examples of intelligent services that can make use of background records [35] in a greenhouse, using methods that reveal information about the ecosystem of their devices and users. Martini et al. [36] using multispectral imaging techniques and the Raspberry Pi, they built a model to predict the amount of clay in the soil. The portability of this type makes it ideal for use in a variety of farm settings where the soil's chemical composition must be evaluated. Farooqui et al. [6]also proposed an IoT-based greenhouse agriculture system and proposed model. The computer models made use of sensors that allow for remote monitoring of agricultural fields and provide advice for improving crop yields based on those observations. Soil chemical content and crop prediction using ML and past data [36]. By automation formerly labor-intensive tasks, these smart models and approaches help farmers increase output while decreasing costs. This encompasses highly precise activities like manure application and irrigation [36]. Knowing how much of each resource to use and when reduces waste and helps preserve natural habitats [37] pollutants as well as simple agricultural monitoring.

Precision agriculture and predictive analytics are two closely related concepts that have gained significant attention in modern agricultural practices. Precision agriculture uses advanced technologies to optimize farming work, while predictive analytics harnesses data-driven models to forecast future events. This section explores the intersection of these two fields, highlighting their applications, benefits, challenges, and potential for transforming agriculture.

9.2.3 Applications of Precision Agriculture System

The integration of precision agriculture and predictive analytics offers several benefits, including increased productivity, resource efficiency, reduced environmental impact, and better risk management. However, challenges include data collection and quality issues, the complexity of implementing technology, and the need for skilled personnel to interpret data and model outputs accurately.

As technology advances, the synergy between precision agriculture and predictive analytics is likely to deepen. Incorporating AI and ML enhances the accuracy of predictive models, enabling real-time decision support systems. Collaborations between agricultural scientists, data analysts, and technology developers will play a crucial role in driving innovation and overcoming challenges.

The combination of precision agriculture and predictive analytics holds immense potential for revolutionizing agricultural practices. By controlling data-driven insights, farmers will be able to manage decisions that optimize resource utilization, increase yields, and promote sustainable farming methods. Continued research and development in these fields will shape the future of agriculture, contributing to global food security and environmental stewardship.

Based on their research, the authors provide various learning models used in intelligent farming. Important farming procedures, such as soil chemical analysis, weather forecasting, predicting yields of crops, and irrigation management models, are all discussed, as is the impact of recent developments in ML on these prediction tasks [27]. Focusing on different parameters related to fertilizers, soil nutrients, and favorable weather conditions for several crops, this study aims to offer a brief overview of recent research on the application of ML techniques in creating intelligent farming models. The application of the precision agricultural system is represented in Figure 9.1.

Precision agriculture utilizes technologies such as GPS, remote sensing, drones, and sensors to collect data on different parameters, and then analyze the data to enable precise decision-making in various agricultural processes such as:

Variable Rate Application: Precision agriculture allows farmers to apply inputs such as fertilizers, pesticides, and water at varying rates across a field. This optimizes resource utilization and minimizes waste.

Yield Monitoring: Advanced sensors and GPS enable real-time yield monitoring, helping farmers identify areas of high and low productivity within a field. This information supports better management strategies.

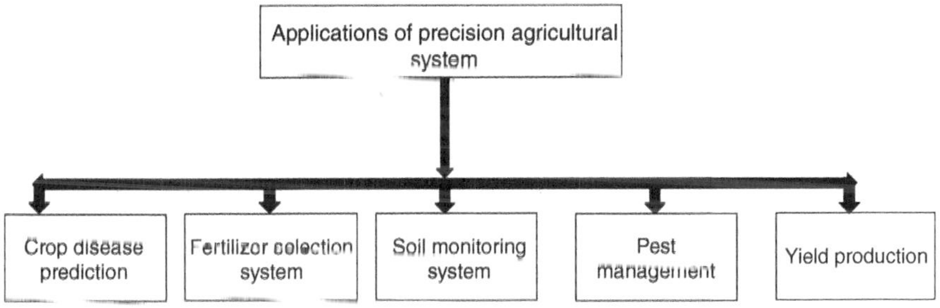

Figure 9.1 Applications related to precision agriculture.

Site-Specific Crop Management: Depending on the soil type, the availability of nutrients, and the stage of development of the crop, producers can adjust their methods to meet the requirements of individual sections of the field.

Pest and Disease Management: Precision agriculture enables early detection of pest and disease outbreaks through remote sensing and predictive models, allowing for targeted interventions.

Soil Health Improvement: Detailed soil mapping and analysis aid in understanding soil composition and structure, facilitating soil health improvement strategies.

9.3 Precision Agriculture Technologies and Methods

To gather and assess all the data needed for precision farming, specialized software and hardware are required. Precision agriculture requires the use of hardware and software, both must be installed by trained professionals. Farming, though, has the capacity on its own with some effort. Precision farming methods can be broken down into three categories: ground, air, and space. The former can be used for machinery management, production planning, mapping, and surveying. The last two are applicable to the resolution of more global challenges, such as everywhere and anytime produce level assessment. Combining different methods can help you get more accurate results.

9.3.1 The Role of GPS in Modern Crop Management

The GPS is a satellite-based navigation system that provides location and time information in all weather conditions, anywhere on the Earth [38]. Figure 9.2 is an example of an image of a satellite-based navigation system based on GPS. In precision agriculture, GPS provides precise location information for mapping and

Figure 9.2 Satellite-based navigation system captured farmland images.

navigation purposes. One of the main ways GPS is used in precision agriculture is for mapping and field-level data collection. Farmers can use GPS receivers mounted on their vehicles or equipment to collect data, shape, and size of the fields, as well as the location of specific features within the fields such as irrigation systems, drainage ditches, and fence lines. The detailed maps of fields are used for planning and scheduling field operations, implementation of variable rate technology in agricultural input management, and analyzing field performance.

Farmers can also use GPS-based guidance systems to help them navigate their vehicles and equipment around the fields with a high intensity of accuracy, reducing the risk of errors and increasing efficiency.

9.3.2 Architecture of a Smart Farming Monitoring System

The need for farming methods that are both effective and environmentally friendly has increased in recent years. Smart farming, also known as precision agriculture, has emerged as a promising solution to address this need. With the help of innovative technologies, such as the IoT, cloud-based computing, and handheld devices, and data analytics, smart farming monitoring systems offer real-time insights and automation capabilities to optimize crop production and resource management [8]. A smart farming monitoring system's design can impact cutting-edge tools that have the potential to transform the agriculture sector. By integrating sensors, communication infrastructure, data analytics, and visualization, these systems enable farmers to make data-driven decisions, optimize resource allocation, and enhance crop yields. The benefits of smart farming systems extend beyond increased productivity, as they also contribute to sustainable resource management and risk mitigation. As the demand for efficient and sustainable agricultural practices continues to rise, the adoption of smart farming monitoring systems is set to become increasingly prevalent in the coming years. The framework for smart farming is shown in Figure 9.3.

There are various components in a smart farming system, but the main components are:

1) **Wireless Sensor Network:** At the heart of a smart farming monitoring system lies a network of sensors that collect data on various environmental parameters such as temperature, humidity, soil moisture, and light intensity. These sensors are strategically placed across the farm to ensure comprehensive coverage [3]. The sensors send their data to a central hub instantaneously so that it may be processed and evaluated.
2) **Communication Infrastructure:** A robust communication infrastructure is essential for seamless data transmission between the sensors, research central hub, and other components of the smart farming system. It includes wireless

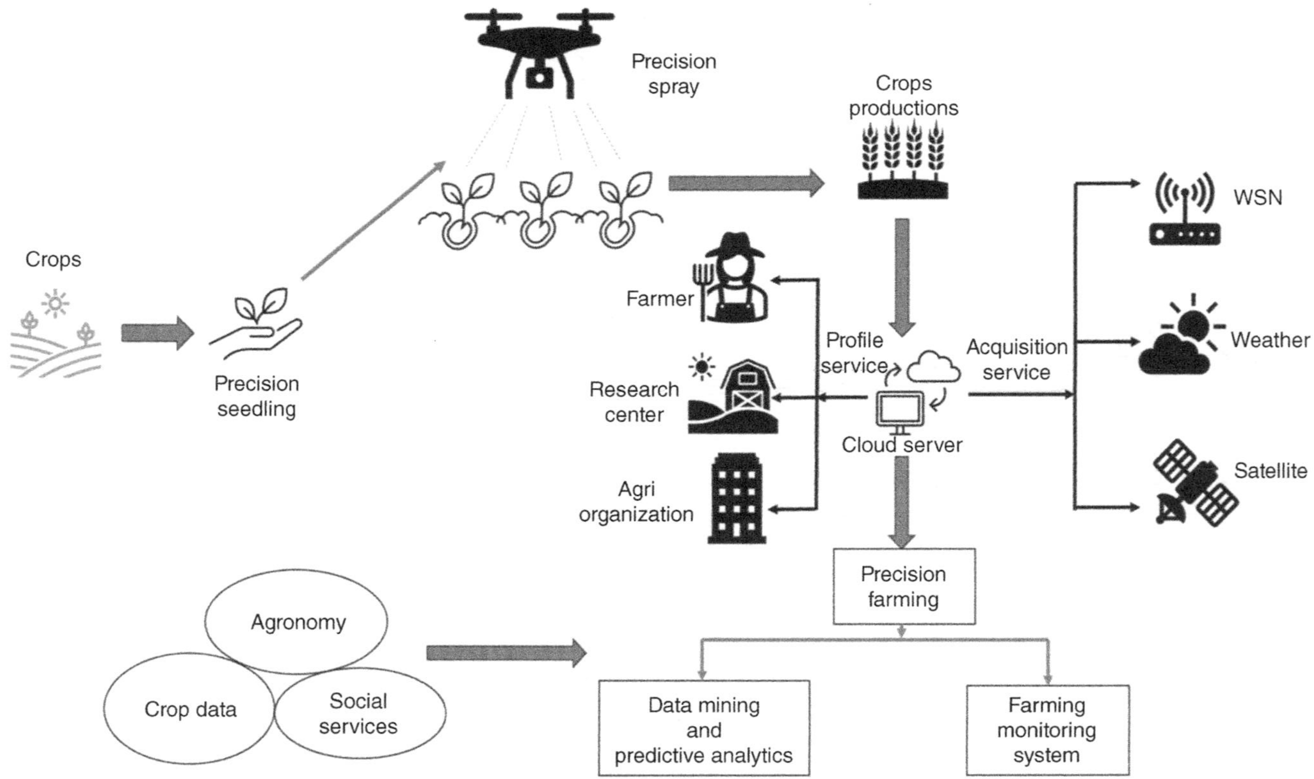

Figure 9.3 Framework for the smart farming system.

protocols and a satellite system to communicate the whole system and maintain the communication efficiently [11].

3) **Research Central Hub:** The research central hub serves as the nerve center of the smart farming monitoring system. It receives data from the sensor network and performs real-time analysis to generate actionable insights. The research central hub can be a physical server or a cloud-based platform that controls cloud computing to handle large data and perform complex analytics algorithms. There are various data scientists who analyze the data and perform evaluations and measurements.
4) **Predictive Analytics:** Applying cutting-edge analytics methods, the information gathered by the network of sensors is processed and studied. ML algorithms are employed to recognize patterns, predict crop development, identify anomalies, and optimize resource allocation. These insights enable agriculturalists to make updated decisions and take proactive measures to enhance yields and minimize resource wastage [3].
5) **Visualization and User Interface:** The output of the data analytics process is presented to the framers through intuitive visualizations and user interfaces. This allows farmers to easily interpret the data and monitor the status of their crops in real time. The user interface can be accessed through web or mobile applications, providing farmers flexibility and convenience.

There are various benefits of the smart farming system to increase the productivity and monitoring the crop and maintain the weed control system which improve the efficiency of production of crops. By continuously monitoring environmental conditions and crop health, smart farming systems enable farmers to optimize resource allocation, such as water, fertilizers, and pesticides. This leads to improved efficiency and reduced costs. Real-time insights provided by smart farming systems help farmers identify potential issues and take timely corrective actions [39]. This leads to better crop management practices, increased crop yields, and superior quality. Smart farming systems promote sustainable agricultural practices by minimizing resource wastage. By precisely applying water and fertilizers based on crop needs, farmers can conserve resources and reduce environmental impact. Smart farming systems provide early detection of diseases, pests, and adverse weather conditions. By quickly identifying these risks, farmers can implement preventive measures, reducing crop losses and ensuring a more stable income.

9.3.3 Agricultural Economic Management

The agriculture system has a profound impact on the global economy due to its role in providing food, raw materials, employment, and trade opportunities. Agriculture plays a major part in the global economy, food security, trade, and overall economic development. The agriculture system is intricately linked to the global economy, affecting food security, employment, trade, raw material supply, rural development, innovation, and more. Changes in agricultural practices, technologies, and policies can have far-reaching consequences for both individual economies and the world economy [34]. Economic management in agriculture is rooted in agricultural production and aims for efficiency in both processes and outputs. Figure 9.4 is a simplified representation of how agricultural economic management in the information period could operate.

Sensors, weather and climate monitoring stations, satellites, and government agencies all contribute to the datasets that are used to improve agricultural economic management [38]. Data collection is crucial since it ensures the continuity of data flows prior to storage. While their intended evolution will cover the transaction service, the focus here is on the components that make up first-rate data and profile services. Many farmers now use precision-based agriculture technology to increase crop production and better manage the global economy thanks to significant developments in big data, data gathering, decision-making systems, and agricultural technological advancements [17].

The sensors in the farming field to improve farm accuracy and economic management by monitoring and reporting on factors such as soil condition, temperature, and air pressure. To enhance the quality of crop yields, deep learning enables the creation of robotic decision-making tools that utilize the appropriate amount of input at the correct time and in the proper locations. With innovative automated irrigation systems, farmers no longer need to keep tabs on agricultural

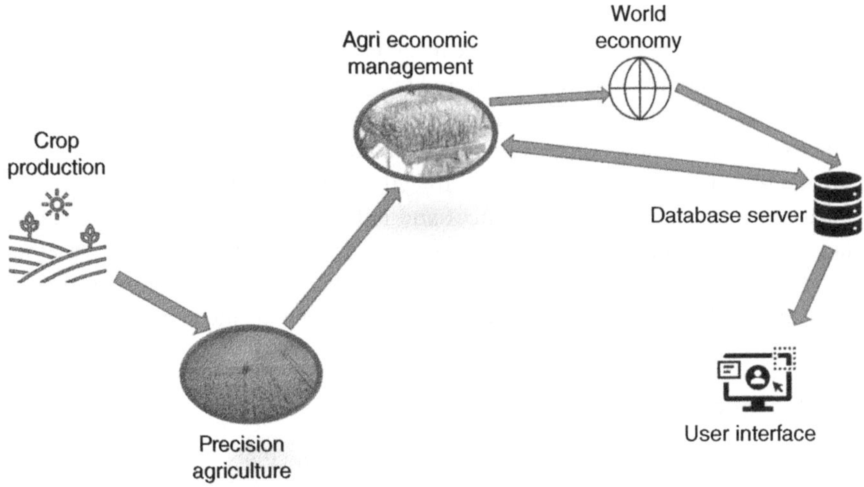

Figure 9.4 Precision agricultural economic management.

economic management to gather data [11]. By using sensors to collect present-time data on temperature, soil, and air quality, the precision agriculture predictive monitoring system may be carried out in a more precise manner that makes use of agricultural economic management. Farmers can save time and effort in continuous field monitoring with the help of the smart precision agriculture system, which employs intelligent protocols and hybrid methods to carry out field observations automatically. To address economic concerns, predictive analysis is necessary to enable automatic irrigation when there is an increase in water demand [6].

9.4 Smart Agriculture Cultivation Recommender System

The goal of the Smart Agriculture Cultivation Recommender System is to help farmers maximize their use of water resources in preparation for more precise farming in the future. Its use in agricultural fields of a certain size ensures efficient use of water and a plentiful harvest. Due to climate change and increased global population, there will be an unprecedented demand for water in the 2050. Finding a way to maximize agricultural water use to satisfy both food and water needs is difficult. Factors associated with crop development are considered to make the most informed farming decisions possible [40]. Providing healthy meals is essential to maintaining human life. In places like Africa, necessities, such as food and water, are in short supply.

Due to environmental uncertainty, the labor and time required by conventional farming methods results in lower yields [41]. The range limitation of existing solutions is still a problem in precision agriculture [42]. Understanding the ecosystem of the cultivated field and strategically allocating water resources are essential for maximum output. Typically, suggestions for soil testing are put off until later [43]. Figure 9.5 depicts the framework of the ACRIS Multiparameter Optimization System for Precision Agriculture. The AMOP method enables farmers to evaluate the current irrigation schedule used throughout crop growth and make decisions based on that evaluation. The precipitation, rainfall, and evaporated water levels schedule irrigation are predicted using crop growth characteristics as input datasets [44].

The state-of-the-art Agro ANN_Genetic Algorithm (AAGA) was used to create the AMOP system, which is a combination of the ANN and the GA. Users can determine the optimal watering schedules for their farms by using the AMOP system built within the cutting-edge AAGA. Through the AMOP system's moderation and integration, predictions of precipitation and evapotranspiration ranges can be made, soil water stability can be initiated through controlled irrigation, and production output can be raised. Since an ANN and a GA have never been combined in such a way before, this is the system's main

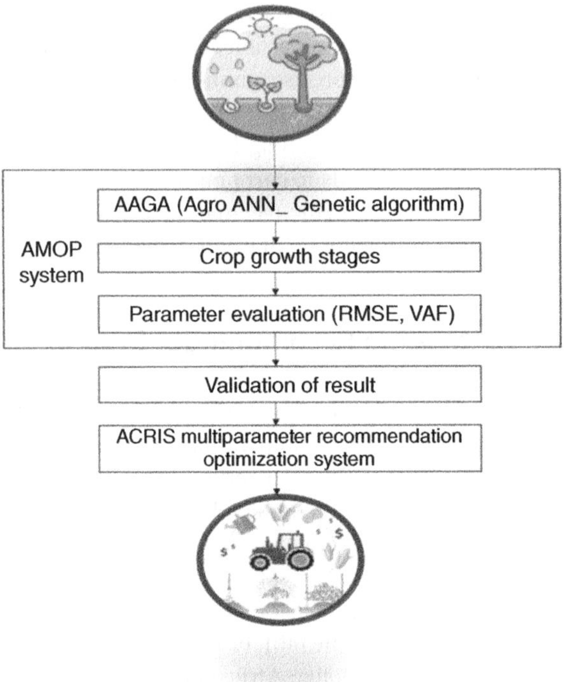

Figure 9.5 Agriculture cultivation recommendation system.

innovation [44]. The AAGA makes use of the water's velocity for irrigation purposes. It had been forecast to rain, albeit adjustments had to be made to the predicted rainfall totals.

9.5 Conclusion

Predictive models assist farmers in making decisions by using climate, soil, air, crop growth, equipment, labor, and availability data. This is precision agriculture. Big data can help precision agriculture acquire massive streaming data in real time. Large data scale requirements challenge data analysis reliability and performance. Precision agriculture is needed to increase farm output and ensure safe, healthy food. Precision agriculture enhances traditional agriculture with data and computers. Precision agriculture in optimizing resource utilization, reducing environmental impact, and improving overall crop yield. They emphasize the role of advanced technologies such as remote sensing, GIS, and ML algorithms in collecting and analyzing vast amounts of data to make accurate predictions and informed decisions. The concepts such as monitoring crops, soil analysis, and watering schedule management are addressed in this in-depth examination of precision farming. It covers topics such as utilizing historical data and real-time monitoring to make precise predictions about crop diseases, fertilizer usage, and yield. We also discuss the challenges and limitations of implementing precision agriculture and predictive analytics, such as data privacy concerns, infrastructure requirements, and the need for skilled professionals. They provide insights into potential solutions and strategies to overcome these barriers, including public–private partnerships and capacity-building initiatives. The Agriculture Cultivation Recommender and Smart Irrigation System is also discussed, and their working and methods used in that system. The overall predictive analytics improve the performance of the recommendation and monitoring system of crops.

References

1 N. R. Council (2010). *Toward Sustainable Agricultural Systems in the 21st Century*. National Academies Press.

2 Khanna, A. and Kaur, S. (2019). Evolution of internet of things (IoT) and its significant impact in the field of precision agriculture. *Computers and Electronics in Agriculture* 157: 218–231.

3 Singh, P.K. and Sharma, A. (2022). An intelligent WSN-UAV-based IoT framework for precision agriculture application. *Computers and Electrical Engineering* 100: 107912.

4 Triantafyllou, A., Sarigiannidis, P., and Bibi, S. (2019). Precision agriculture: a remote sensing monitoring system architecture. *Information* 10 (11): 348.

5 Hassler, S.C. and Baysal-Gurel, F. (2019). Unmanned aircraft system (UAS) technology and applications in agriculture. *Agronomy* 9 (10): 618.

6 Farooqui, N.A., Mishra, A.K., and Mehra, R. (2022). IOT based automated greenhouse using machine learning approach. *International Journal of Intelligent Systems and Applications in Engineering* 10 (2): 226–231.

7 Hakkim, V.M.A., Joseph, E.A., Gokul, A.J.A., and Mufeedha, K. (2016). Precision farming: the future of Indian agriculture. *Journal of Applied Biology and Biotechnology* 4 (6): 68–72.

8 Liang, C. and Shah, T. (2023). IoT in agriculture: the future of precision monitoring and data-driven farming. *Eigenpub Review of Science and Technology* 7 (1): 85–104.

9 Firouraghi, N., Kiani, B., Jafari, H.T. et al. (2022). The role of geographic information system and global positioning system in dementia care and research: a scoping review. *International Journal of Health Geographics* 21 (1): 1–13.

10 Ahmed, W., Zhang, Q., Lobos, A. et al. (2018). Precipitation influences pathogenic bacteria and antibiotic resistance gene abundance in storm drain outfalls in coastal sub-tropical waters. *Environment International* 116: 308–318.

11 Muangprathub, J., Boonnam, N., Kajornkasirat, S. et al. (2019). IoT and agriculture data analysis for smart farm. *Computers and Electronics in Agriculture* 156: 467–474.

12 Oliver, S.T., González-Pérez, A., and Guijarro, J.H. (2019). Adapting models to warn fungal diseases in vineyards using in-field internet of things (IoT) nodes. *Sustainability* 11 (2): 416.

13 Elkarim, G.A., Alotaibi, N.M., Samuel, N. et al. (2017). Social media networking in pediatric hydrocephalus: a point-prevalence analysis of utilization. *Journal of Neurosurgery Pediatrics* 20 (2): 119–124.

14 Jawad, H.M., Nordin, R., Gharghan, S.K. et al. (2017). Energy-efficient wireless sensor networks for precision agriculture: a review. *Sensors* 17 (8): 1781.

15 Sikeridis, D., Tsiropoulou, E.E., Devetsikiotis, M., and Papavassiliou, S. (2018). Wireless powered public safety IoT: a UAV-assisted adaptive-learning approach towards energy efficiency. *Journal of Network and Computer Applications* 123: 69–79.

16 Hsu, T.-C., Chang, S.-C., and Hung, Y.-T. (2018). How to learn and how to teach computational thinking: suggestions based on a review of the literature. *Computers and Education* 126: 296–310.

17 Rehman, A., Saba, T., Kashif, M. et al. (2022). A revisit of internet of things technologies for monitoring and control strategies in smart agriculture. *Agronomy* 12 (1): 127.

18 Sinha, B.B. and Dhanalakshmi, R. (2022). Recent advancements and challenges of internet of things in smart agriculture: a survey. *Future Generation Computer Systems* 126: 169–184.

19 Farooqui, N.A., Mishra, A.K., and Mehra, R. (2022). Concatenated deep features with modified LSTM for enhanced crop disease classification. *International Journal of Intelligent Robotics and Applications* 1–25.

20 Ramesh, P.S., Kokila, T., and Geetha, D. (2015). Plant mediated green synthesis and antibacterial activity of silver nanoparticles using *Emblica officinalis* fruit extract. *Spectrochimica Acta, Part A, Molecular and Biomolecular Spectroscopy* 142: 339–343.

21 Farooqui, N.A., Mishra, A.K., and Mehra, R. (2022). Automatic crop disease recognition by improved abnormality segmentation along with heuristic-based concatenated deep learning model. *Intelligent Decision Technologies* 16 (2): 407–429.

22 Cetişli, B. and Büyükçingir, E. (2013). Time series prediction of apple scab using meteorological measurements. *African Journal of Biotechnology* 12 (35): 5444–5451.

23 Balducci, F., Impedovo, D., and Pirlo, G. (2018). Machine learning applications on agricultural datasets for smart farm enhancement. *Machines* 6 (3): 38.

24 Hamad, A.F., Han, J.-H., Kim, B.-C., and Rather, I.A. (2018). The intertwine of nanotechnology with the food industry. *Saudi Journal of Biological Sciences* 25 (1): 27–30.

25 Kan, Z.-R., Liu, Q.-Y., He, C., and Jing, Z.-H. (2020). Responses of grain yield and water use efficiency of winter wheat to tillage in the North China Plain. *Field Crops Research* 249: 107760.

26 Kurşun, G. and Dengiz, O. (2020). Assessment of land suitability for the production of major crops in Ayrancı district of Karaman province located at arid terrestrial ecosystem. *Eurasian Journal of Soil Science* 9 (1): 24–33.

27 Farooqui, N.A. and Mehra, R. (2020). A machine learning approach to simulating farmers' crop choices for drought prone areas. *Proceedings of ICETIT 2019: Emerging Trends in Information Technology*, New Delhi, India (21–22 June 2019). Springer, pp. 472–481.

28 Adamides, G., Kalatzis, N., Stylianou, A. et al. (2020). Smart farming techniques for climate change adaptation in Cyprus. *Atmosphere (Basel)* 11 (6): 557.

29 Balafoutis, A., Beck, B., Fountas, S. et al. (2017). Precision agriculture technologies positively contributing to GHG emissions mitigation, farm productivity and economics. *Sustainability* 9 (8): 1339.

30 Borlaug, N.E. (2002). *The Green Revolution Revisited and the Road Ahead*. Stockholm: Nobelprize.org.

31 Ali, M.S., Miah, M.S., Haque, J. et al. (2021). An enhanced technique of skin cancer classification using deep convolutional neural network with transfer learning models. *Machine Learning with Applications* 5: 100036.

32 Pandiarajaa, P. (2021). A survey on machine learning and text processing for pesticides and fertilizer prediction. *Turkish Journal of Computer and Mathematics Education (TURCOMAT)* 12 (2): 2295–2302.

33 Parikh, N., Pandey, M., Kumar, M. et al. (2022). Identifying dominant recombination mechanisms in spiro-based conventional perovskite solar cells: roles of interface and bulk recombination. *Energy Reports* 8: 7957–7963.

34 Jat, R.A., Jain, N.K., Yadav, R.S. et al. (2023). System-based integrated nutrient management improves productivity, profitability, energy use efficiency and soil quality in peanut wheat cropping sequence in light black soils. *Sustainability* 15 (2): 1361.

35 Migliavacca, M., Musavi, T., Mahecha, M.D. et al. (2021). The three major axes of terrestrial ecosystem function. *Nature* 598 (7881): 468–472.

36 Martini, B.G., Helfer, G.A., Barbosa, J. et al. (2021). IndoorPlant: a model for intelligent services in indoor agriculture based on context historics. *Sensors* 21 (5): 1631.

37 Karthik, M., Abhinav, J., and Shankar, K.V. (2021). Morphological and mechanical behaviour of Cu–Sn alloys—a review. *Metals and Materials International* 27: 1915–1946.

38 Mehra, R., Farooqui, N.A., and Tyagi, A. (2020). Data mining and fusion techniques for wireless intelligent sensor networks. In: *Handbook of Wireless Sensor Networks: Issues and Challenges in Current Scenario's*, vol. 1132 (ed. P.K. Singh, B.K. Bhargava, and M. Paprzycki). Cham: Springer https://doi.org/10.1007/978-3-030-40305-8_28.

39 Qazi, S., Khawaja, B.A., and Farooq, Q.U. (2022). IoT-equipped and AI-enabled next generation smart agriculture: a critical review, current challenges and future trends. *IEEE Access* 10: 21219–21235.

40 Partarini, N.M.C. (2019). Validasi Data Curah Hujan TRMM (Tropical Rainfall Measurement Mission). In: *Sebagai Alternatif Data Hidrologi di Sub-DAS Lesti'* (ed. N.M.C. Partarini), 1–214. Universitas Brawijaya.

41 Xia, G.-S. Bai, X., Ding, J. *et al.* (2018). DOTA: a large-scale dataset for object detection in aerial images. *Proceedings of the IEEE Conference on Computer Vision and Pattern Recognition*, pp. 3974–3983, Salt Lake City, UT (18–23 June 2018). New York: IEEE.

42 Ilie-Ablachim, D., Pătru, G.C., Florea, I.-M., and Rosner, D. (2016). 'Monitoring device for culture substrate growth parameters for precision agriculture: Acronym: MoniSen. *2016 15th RoEduNet Conference: Networking in Education and Research*, pp. 1–7, Bucharest, Romania (7–9 September 2016). New York: IEEE.

43 Sureshkumar, S., Kirthiga, N., Kumar, T.A. et al. (2023). Dual access control for cloud-based data storage and sharing. *2023 2nd International Conference on Vision Towards Emerging Trends in Communication and Networking Technologies (ViTECoN)*, Vellore, Tamil Nadu, India (5–6 May 2023). New York: IEEE, pp. 1–6.

44 Kumar, P., Udayakumar, A., Kumar, A.A. et al. (2023). Multiparameter optimization system with DCNN in precision agriculture for advanced irrigation planning and scheduling based on soil moisture estimation. *Environmental Monitoring and Assessment* 195 (1): 13.

10

A Simple Way to Comprehend the Difference and the Significance of Artificial Intelligence in Agriculture

Karan Aggarwal[1], Ruchi Doshi[2], Maad M. Mijwil[3], Kamal Kant Hiran[4], Murat Gök[5], and Indu Bala[6]

[1] *Electronics and Communication Engineering Department, Maharishi Markandeshwar (Deemed to be University), Ambala, India*
[2] *Department of Computer Science and Engineering, Universidad Azteca, Chalco, Mexico*
[3] *Computer Techniques Engineering Department, Baghdad College of Economic Sciences University, Baghdad, Iraq*
[4] *School of Computer Science and IT, Symbiosis University of Applied Sciences, Indore, India*
[5] *Department of Computer Engineering, Yalova University, Yalova, Turkey*
[6] *School of Electrical and Electronics Engineering, Lovely Professional University, Punjab, India*

10.1 Introduction

At this time, the devices technology revolution is developing day by day. We live in the era of devices that help us accomplish many tasks in our lives. Technology allows solving people's economic situations unprecedentedly that they would not have dreamed of 20 years ago. In computer science, the science of artificial intelligence (AI) appeared, which is an essential part of it, and also the science that is more developed and applied in many fields, including the medical domain, military, people tracking, social networking sites, and many endless applications [1, 2]. The science of AI can be represented as the ability of a machine or a computer to execute tasks related to the human mind, such as thinking, reasoning, generalization, and learning from past things and experiences to perform functions that are greater than or equal to the ability of humans to accomplish [3]. In 1956, a concept appeared that changed many ideas in computer science which is AI, which was mentioned at the Dartmouth conference. In this event, we discussed how to realize and support computer software intelligently. The term AI did not meet the purpose of some of the scientists present, including Newell and Simon,

Intelligent Techniques for Predictive Data Analytics, First Edition.
Edited by Neha Singh, Shilpi Birla, Mohd Dilshad Ansari, and Neeraj Kumar Shukla.

Published 2024 by John Wiley & Sons, Inc.

and at the same time, John McCarthy, Marvin Minsky, and Claude Shannon, who helped a lot in building the idea of AI, approved it. The idea of creating robots is not new, as some think. On the contrary, it is an ancient idea that dates back many years. Nevertheless, examples exist of the concept of giving life to inanimate objects and transforming them into intelligent and efficient objects. In Greek myths, Daedalus' attempt to produce artificial humans who ruled the wind can be considered the first example of AI. In general, scientists always seek to develop machines based on the idea of AI by linking computers with intelligence [4, 5]. The most significant example of this is the Turing machine (*the enigma machine*) in 1943 by the English mathematician Alan Turing.

Turing claimed in 1950, in his famous article "Computing Machinery and Intelligence", that computers could calculate and develop a test to decide whether or not a machine is intelligent [6]. After 1960, many studies described the difference between the human brain and the computer's ability to control things. On this basis, engineering and cognitive modeling were designed and narrowed the gap between man and machine, which is still continuing until now. However, man is the engineer of machines and lays the basis for their assignment and programming in the way they wish. But the most critical question is whether computers will have the ability to think uniquely like humans and operate without the intervention of any human hand. In other words, can human consciousness as a biological being be transferred to a machine? There are many inquiries about the future of AI technologies [7].

Today, we are witnessing a lot of articles and research that talk about the integration of computers into our future lives [8, 9]. Also, AI has a significant role in confronting diseases, especially the COVID-19 pandemic, as it is able to track and diagnose computerized images (Ct-scan) and chest X-ray images [10–12]. Many science fiction films deal with this idea of integrating computers into human life, the most famous of which are *Transcendence*-2014, *Her*-2013, *A.I. AI*- 2001, *Ex Machina*-2014. In this work, we will briefly and usefully address the three most significant topics that are widely circulated and applied in many fields: machine learning and deep learning, which are considered a large part of AI, also data science.

Evolutions in AI are somewhat complex and can be summarized in two concepts: machine learning and deep learning. These terms sometimes sound like buzzwords which are actually synonyms, which is why it is essential to know how to differentiate them and why they are so important in use. There are a lot of examples around; for instance, Netflix knows what movies and series TV you like to watch, Facebook knows who shows a destination in the photo, and both are reasons for self-driving cars (Autonomous) [13]. A customer service representative knows if you are satisfied with their service even before you fill out a customer satisfaction survey. This article focuses on finding out the difference between machine learning, deep learning, and data science.

10.2 Machine Learning

It is considered the most influential branch of AI and is responsible for designing algorithms that are able to learn and do not have to be explicitly programmed. In other words, the developer or programmer will not have to sit for hours to write the code or possible scenarios to produce updated algorithms. All that is required of him/her is to feed these algorithms with vast amounts of data so that the algorithms learn and know what they have to do and accomplish without any concern. There are two types of learning: supervised and unsupervised (see Figure 10.1).

Moreover, making a model in machine learning means designing a mathematical representation by generalizing and learning from the training data. Then, the designed machine learning model is involved in the new data to make predictions and earn effects. The model designed can be a regression model or a classification model based on the target variable known as the Y variable. Therefore, we must create a regression model if the target variable has a quantitative value. In contrast, if the data type of the target variable is qualitative, a classification model must be made. The absence of data can be a challenge when making machine learning models. The most vital metric of a classification machine learning model is accuracy. Besides, precision is the ratio of the total number of proper predictions to the total number of data points in the test dataset. In other words, accuracy is one way to evaluate a model's execution, but there are other ways. Nevertheless, accuracy offers the most reasonable perspective of a model's performance for a given dataset compared to different metrics. We employ metrics to evaluate the execution of a model. A model parameter is a value learned and evaluated during training from the dataset. The value must be approximated from the training data. Accordingly, the model parameters are internal variables of the machine learning model. Its values are set based on training. Machine learning models use these values to create predictions. It is based on the accuracy of model

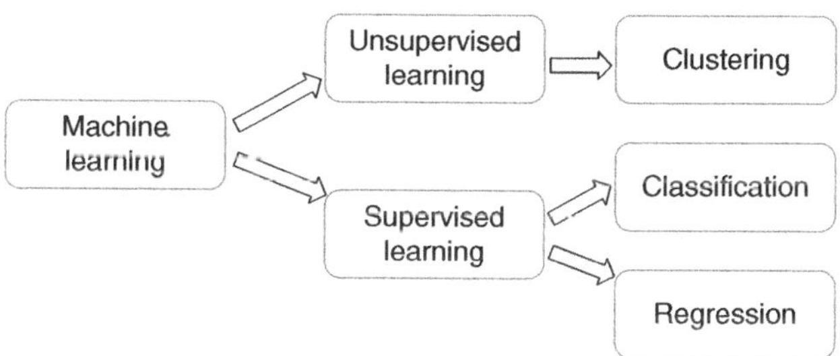

Figure 10.1 Machine learning types. *Source:* Tkatek, Said et al. (2020)/Putting the world back to work: An expert system using big data and artificial intelligence in combating the spread of COVID-19 and similar contagious diseases.

parameter values. Model parameter values have some degree of control over the sophistication of a model. The model complexity curve is a graphical texture employed to express the complexity of a model. It exhibits a relative increase in information as the example size increases. Also, it can be utilized for data deletion and speed up the learning process without concerning classification accuracy.

10.3 Deep Learning

Deep learning is a machine learning method applied in many domains as it is characterized by using many layers of nonlinear processing units to extract and transform features. In addition, each next layer takes the output from the previous layer as an input. Algorithms can be moderated (such as classification) or unsupervised (such as pattern analysis) [14–16]. Deep learning is an evolution machine learning model that utilizes a programmable neural network that allows machines to make sound judgments without human intervention. Table 10.1 illustrates the distinctions between machine learning and deep learning. Deep learning contains a set of architectures that rely heavily on learning and data representation [17]. It supports the design of a hierarchical representation of data by deriving higher-level properties from lower-level properties. This representation learns multiple levels of representation that correspond to different various of abstraction. Moreover, the deep learning model depends mainly on learning from data representation. For example, when a representation of an image is mentioned, it can be considered as a vector of intensity values for each pixel or features such as edge sets and custom shapes. However, some of these features represent the data better. As a benefit of this scene, the deep learning model utilizes efficient algorithms to

Table 10.1 The distinctions between machine learning and deep learning.

Distinctions	Machine learning	Deep learning
Dataset	Small to medium dataset	Big dataset
H/W requirements	Low-end machines	Requires machines with GPU
Engineering peculiarities	Understand the features of the data	Understand the functionality of the data
Training time	Short	Long
Processing time	Seconds or hours	Hours or weeks
Number of algorithm	Many	Few
Data interpretation	Easy and some algorithms are hardly	Difficult

extract the hierarchical features that sufficiently represent the data rather than the handcrafted features. The deep learning model is leveraged to analyze data utilizing a logical structure that draws human-like inferences. In fact, deep learning applications employ an artificial neural network to perform a specific action. The vision of artificial neural networks is inspired by the biological neural network of the human brain, which generates a learning process with much greater potential than those of traditional machine learning models.

Deep learning concentrates on enhancing the AI process, which is where machines learn things. The core of deep learning lies in computers fast enough and enough data to train large neural networks. As a result, many companies employ deep learning and advanced AI to solve product issues and services. Some people say that deep learning is just another term for machine learning, but that's not proper. Deep learning is a subset of machine learning. People should stop trying to make machine learning/deep learning the solution to issues that could be solved more efficiently with straightforward math. Machine learning techniques have been utilized for a long time, but deep learning is far superior to its peers. Deep learning is effectively a way to perform pattern matching on a large scale. More significantly, deep learning has only had limited success in particular areas. Some specialists believe that reinforcement learning involves the development of AI models without providing them with an enormous amount of data labels. Deep reinforcement learning is one of the most exciting areas of AI investigation; it has limited sensation in solving real-world issues.

10.4 Data Science

Data science can be expressed as a practice that aims to deal with all forms of data in terms of gathering, coordinating, analyzing, reviewing, and extracting the most critical features. This science integrates computer science fields and statistics with other modern sciences. Nowadays, data science is one of the most requested disciplines in the labor market because it is the basis on which many disciplines such as machine learning, deep learning and big data are based. It is the science that mainly supports the development of AI techniques. Companies seek to use data science techniques in analyzing their data and making decisions that help in expanding the business more broadly. Making decisions is not straightforward, so data science is a mixed and multidisciplinary education that includes mathematics, computer science, statistics, and business. The process of evidence analysis helps improve the quality of commercial, governmental, and administrative decision-making and scientific research. The development of data science is being leveraged in many fields like the medical, military, agriculture and more. Data science provides significant insights into many complex matters, in many circumstances, with incredible accuracy and timeliness. Data science includes many vital and more complex processes that seek to predict, classify, or identify a pattern in the data or generate recommendations before taking a unique action. Data science is an interdisciplinary approach to finding and extracting multiple types of data by integrating analytical methods, specialist expertise, and modern technologies. This approach usually relies on the areas of data mining and forecasting using machine learning techniques, analysis of such predictions, statistics, and text analysis. In general, most organizations are short of experts to transform their big data to find insights and troubleshoot problems the organization is unaware of. To properly understand and monetize the value of data science, organizations will need to include forecasting in their data and improvement in their operational and business systems. Nowadays, many institutions seek to train specialized cadres in analyzing the data of their systems, who will contribute to the development and implementation of projects through machine learning techniques. In short, this field is in constant development, and we may hear a lot about it in the coming years. Data science includes illustrative, diagnostic, predictive, and educational capabilities, which indicates that with data science, organizations can employ data to learn what occurred, why it occurred, what will occur, and what they should do with the expected effect.

10.5 AI in the Agriculture Industry

10.5.1 Techniques in Agriculture

In this section, the significance of AI techniques in the agricultural industry and how modern technology contributes to the development of the agricultural sectors are reviewed. In this sector, many food producers suffer from threats and risks to their crops and the emergence of a group of diseases that damage crops. In general, the agricultural sector depends on natural forces in managing crops. In other words, the change in the nature of the climate and the lack of rainfall led to a shortage of workers available to work in this sector, which led to a decrease in agricultural production, especially in Iraq, which is considered one of the countries most affected by climate change and the lack of crops and the heavy dependence on imports from neighboring nations, which led to a significant increase in food prices. With all these challenges, the main issue that preoccupies the countries of the world is automation in the agricultural industry. The population is growing exponentially daily, and with the increase in population density, the necessity for food and work also increases daily. The methods used by agriculturalists are not sufficient to meet these requirements. Therefore, modern automated methods have been added to make things more straightforward and more productive. With the growth of AI techniques, applying this science in the agricultural sector has become vital. Scholars worldwide seek to conduct a set of research on using AI techniques in the agricultural industry. In the coming years, AI will have sufficient capacity to produce better predictive insights, as it will contribute significantly to increasing agricultural production while providing job opportunities for millions of people worldwide [18, 19].

There is a lot of literature in which deep learning techniques have been utilized in diagnosing and classifying crop diseases. In a study by Tiwari et al. [20], they advised applying a deep learning technique like the VGG-19 model to investigate and detect potato leaf diseases. This study got an accuracy effect of more than 97%. In another study by Rangarajan et al. [21], they applied two deep learning-based architectures, AlexNet and VGG-16 net, to classify tomato crop disease. This study got the accuracy effects of both architectures at 97.29% and 97.49%, respectively. The agricultural industry can benefit from solutions powered by AI by managing environmental challenges using weather forecasts, establishing a soil and crop health monitoring system, and predictive analysis through the use of techniques that help collect information and data needed to make decisions about how to enhance production and also creating designs to detect rusts harmful to agriculture. In recent times, machine learning and deep learning techniques have become essential for the agricultural industry to rely on them to improve efficiency and production continuously. Moreover, these techniques help agriculturalists predict demand by analyzing information, knowing the population growth rate, and finding new areas that can be exploited to establish agricultural lands to keep pace with this population growth. Furthermore, these techniques help in the process of spraying and removing poison weeds and useless weeds through the use of drones (see Figure 10.2). Finally, these techniques help save excessive use of pesticides, water, and herbicides, maintain soil fertility, help in efficient use of labor and increase productivity while enhancing quality.

10.5.2 Smart Agriculture

In recent years, many governments have started to develop and apply smart agriculture, which refers to the application of modern and advanced technologies in the agricultural sector, as it comes to integrating the latest technological innovations with the aim of improving resources and making the most suitable decisions by obtaining accurate information and data about the state of agricultural products and the weather. Smart agriculture has become a significant trend that is rapidly developing thanks to the developments that help agriculturalists establish an advanced agricultural field, which is called precision agriculture. In addition, smart agriculture contributes to solving many of the problems faced by agriculturalists by relying on intelligent applications that contribute to increasing

Figure 10.2 Drones in agriculture sector. (a) Planting drone. (b) Irrigation drone. (c) Soil analysis drone. (d) Crop monitoring drone. (e) Crop spraying drone. (f) Health assessment drone. *Source:* From Casadiego [22], ORCID, CC0 1.0.

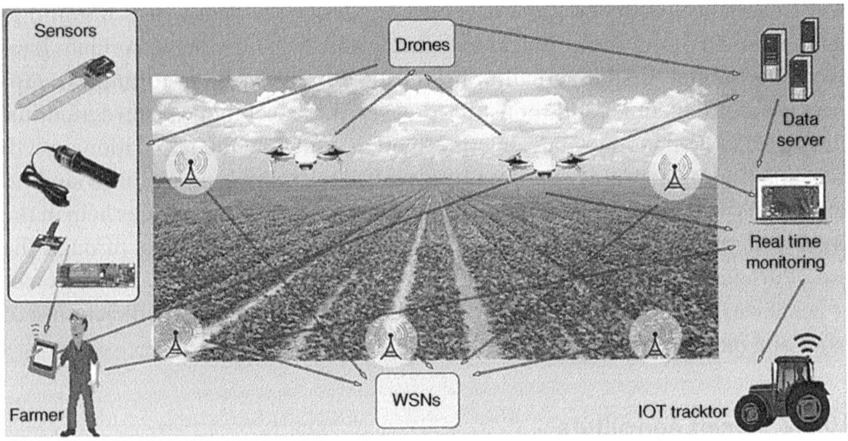

Figure 10.3 Drones send agricultural data to agriculturalists' phones. *Source:* From Rehman et al. [24] / MDPI / CC BY 4.0.

agricultural production by 70%, as expected in 2025 [23]. This percentage may increase beyond 2030. Figure 10.3 illustrates the use of drones to send agricultural data to agriculturalists' phones. Smart agriculture brings many benefits, some of which can be highlighted:

- It contributes to increasing gains as new technologies allow to reduce the cost of resources such as water, energy or fertilizers and increase production with a higher quality that satisfies olives.

- Full control of all planting-related tasks and assist agriculturalists in completing duties in less time while supporting them make the right decisions.
- It contributes to improving the quality of life for farmers and people working in the agricultural sector by integrating tools and applications and facilitating the process of control, incredibly remote control, without needing to be present in the field.
- Smart applications are characterized by the automation of many aspects of the agricultural sector and agricultural production, as they save time and costs.
- Agricultural resolutions based on AI are highly sustainable, as these applications seek to make more sustainable and highly efficient operations.

Big data is one of the things that contribute to the growth of the agricultural sector with AI. Moreover, agriculture has begun to integrate advanced applications that allow the extraction of data, statistics, and numbers increasingly, by extracting effects and determining the type of agriculture that is most productive and in demand by the consumer to meet future needs. Where smart sensors are employed to obtain soil, water, weather, or plant data and deliver these details to the grower's phone efficiently, with graphs illustrating daily agricultural activity, the use of robots and drones is expected to increase by 2030 in agricultural tasks to extract high-quality data with high accuracy and trustworthiness (see Figure 10.4). These drones help in the continuous monitoring of agricultural crops and the detection of potential problems in less time and lower costs, such as controlling pests and preventing the spread of diseases among plants. On the one hand, AI contributes to the creation of applications that allow remote integration of the various elements in the agricultural sector, advanced software to govern and manage the agricultural production process, with the significant development of agricultural machines that employ the global positioning system (GPS) and can be handled through a smartphone. Automation is considered one of the main elements in the process of transition to smart agriculture, where various aspects and elements are automated during the stages of agricultural production based on previous data and records that are made by tools such as smart sensors. For illustration, the automation of irrigation or the weather allows the achievement of suitable climatic conditions for the development of plant production. AI has a significant and influential role in the agricultural sector, as it continuously seeks to develop machines and

Figure 10.4 Involve robots in agricultural production. (a) sompong_tom/Adobe Stock Photos. (b) kinwun/Adobe Stock Photos. (c) Pugun & Photo Studio/Adobe Stock Photos.

tools that help collect crops, predict weather conditions, and provide more efficient systems. Precision agriculture is an AI solution in the agricultural industry. This technology is applied in the health monitoring of plants, field harvesting, weed and pest control, detection of soil shortages, and other tasks that benefit the agriculturalist. Without a doubt, smart agriculture still has a lot to do to reach a very advanced stage and contribute to facilitating decision-making and improving the quality of agricultural crops and the use of smart sensors to monitor changes in climate, water, or soil and control them remotely through a smartphone or laptop or from any other device connected to the internet.

Smart agriculture involves information and data technologies to optimize complicated agricultural systems. The priority is data access and how producers can employ the information organized intelligently. The technology employed in smart agriculture ranges from IoT and robotics to drones and AI. With these implements, agriculturalists can monitor farm essentials without needing on-site physical existence. This allows them to make decisions for the entire area, a lot, or even a single plant. However, smart agriculture is not just for large agriculturalists but can also be utilized by small agriculturists, organic growers, or other smaller-scale operations. The entire smart agriculture technique is handled by software and monitored by sensors. In addition, automation has allowed influential gains in production efficiency, quality gains, and sustainability. One of the challenges of smart agriculture is the interoperability between all the techniques and the different digital initiatives to encourage access to the data generated. As the complexity of agriculture increases, so does the number of parties interested and the sophistication of information flows. Current production techniques employ machines and software from multiple manufacturers. As a result, producers must deal with many proprietary data formats, code lists, etc. Thus, standardized interfaces and data formats are needed to gain interoperability. This interconnection suggests information interactions between manufacturers of inputs for production, distributors, producers, the food industry, and service providers. Also, even clients are included, who increasingly make purchasing decisions based on the availability of such information. In general, the purpose of employing information and communication technology, which is based on getting data, analyzing it, and utilizing it to optimize processes and decision-making, is to enhance the economic effects for the producer, reduce the risks of a specific activity, and encourage increasingly sustainable techniques.

10.6 Conclusions

AI techniques have a significant and distinct role in many different industries, such as transportation, finance, health care, and now agriculture. In addition, it has helped agriculturalists monitor their agricultural products with considerable advantages in spreading their crops. Moreover, AI in agriculture not only helps agriculturalists to automate their cultivation but also leads to precision farming for better crop yield and more excellent quality while using fewer resources. AI employs modern technologies that help agriculturalists monitor their crops without needing individual farm monitoring. Many companies and startups today expect to develop AI in agriculture. AI in agriculture is redefining the traditional pattern of farming. The future of AI in agriculture has become progressive in developing overall transformation using various advanced strategies.

References

1. Dadhich, M., Hiran, K.K., and Rao, S.S. (2021). Teaching–learning perception toward blended e-learning portals during pandemic lockdown. *Advances in Intelligent Systems and Computing* 119–129. https://doi.org/10.1007/978-981-16-1696-9_11.
2. Mijwil, M.M., Gök, M., Doshi, R. et al. (2023). Utilizing artificial intelligence techniques to improve the performance of wireless nodes. In: *Applications of Artificial Intelligence in Wireless Communication Systems* (ed. K. Kumar), 150–162. https://doi.org/10.4018/978-1-6684-7348-1.ch010.
3. Stahl, B.C., Andreou, A., Brey, P. et al. (2021). Artificial intelligence for human flourishing – beyond principles for machine learning. *Journal of Business Research* 124: 374–388. https://doi.org/10.1016/j.jbusres.2020.11.030.
4. Al-mashhadani, M.I., Hussein, K.M., Khudir, E.T., and Ilyas, M. (2022). Sentiment analysis using optimized feature sets in different Facebook/Twitter dataset domains using big data. *Iraqi Journal for Computer Science and Mathematics* 3 (1): 64–70. https://doi.org/10.52866/ijcsm.2022.01.01.007.
5. Salem, I.E., Mijwil, M.M., Abdulqader, A.W., and Ismaeel, M.M. (2022). Flight-schedule using Dijkstra's algorithm with comparison of routes finding. *International Journal of Electrical and Computer Engineering* 12 (2): 1675–1682. https://doi.org/10.11591/ijece.v12i2.pp1675-1682.
6. Turing, A.M. (1950). Computing machinery and intelligence. *Mind* 59 (236): 433–460. https://doi.org/10.1093/mind/LIX.236.433.
7. Verma, S., Sharma, R., Deb, S., and Maitra, D. (2021). Artificial intelligence in marketing: systematic review and future research direction. *International Journal of Information Management Data Insights* 1 (1): 100002. https://doi.org/10.1016/j.jjimei.2020.100002.
8. Zhang, H., Lee, I., Ali, S. et al. (2022). Integrating ethics and career futures with technical learning to promote AI literacy for middle school students: an exploratory study. *International Journal of Artificial Intelligence in Education* 2022: 1–35. https://doi.org/10.1007/s40593-022-00293-3.

9 Hiran, K.K. and Doshi, R. (2013). An artificial neural network approach for brain tumor detection using digital image segmentation. *International Journal of Emerging Trends & Technology in Computer Science* 2 (5): 227–231.

10 Mijwil, M.M. (2023). Deep convolutional neural network architecture to detection COVID-19 from chest X-ray images. *Iraqi Journal of Science* 64 (5): 2561–2574. https://doi.org/10.24996/ijs.2023.64.5.38.

11 Mijwil, M.M. and Al-Zubaidi, E.A. (2021). Medical image classification for coronavirus disease (COVID-19) using convolutional neural networks. *Iraqi Journal of Science* 62 (8): 2740–2747. https://doi.org/10.24996/ijs.2021.62.8.27.

12 Dadhich, M., Hiran, K.K., Rao, S.S., and Sharma, R. (2022). Impact of covid-19 on teaching-learning perception of faculties and students of higher education in Indian purview. *Journal of Mobile Multimedia* 18 (4): 957–980. https://doi.org/10.13052/jmm1550-4646.1841.

13 Ali, A.H., Abdullah, M.Z., Abdul-wahab, S.N., and Alsajri, M. (2020). A brief review of big data analytics based on machine learning. *Iraqi Journal for Computer Science and Mathematics* 1 (2): 13–15. https://doi.org/10.52866/ijcsm.2020.01.01.002.

14 Faieq, A.K. and Mijwil, M.M. (2022). Prediction of heart diseases utilising support vector machine and artificial neural network. *Indonesian Journal of Electrical Engineering and Computer Science* 26 (1): 374–380. https://doi.org/10.11591/ijeecs.v26.i1.pp374-380.

15 Aggarwal, K., Bhamrah, M.S., and Ryait, H.S. (2018). Texture analysis of ultrasound images of liver cirrhosis through new indexes. In: *Studies in Computational Intelligence*, vol. 713, 93–101.

16 Dadhich, S., Pathak, V., Mittal, R., and Doshi, R. (2021). Machine learning for weather forecasting. In: *Machine Learning for Sustainable Development* (ed. K.H. Hiran, D. Khazanchi, A.K. Vyas, and S. Padmanaban), 161–174. De Gruyter https://doi.org/10.1515/9783110702514-010.

17 Lusci, A., Pollastri, G., and Baldi, P. (2013). Deep architectures and deep learning in chemoinformatics: the prediction of aqueous solubility for drug-like molecules. *Journal of Chemical Information and Modeling* 53 (7): 1563–1575. https://doi.org/10.1021/ci400187y.

18 Koyuncu, B. and Gok, M. (2019). Bitki islahinda genotip verim degerinin regresyon yontemleri ile tahmini. *Bahce* 48 (2): 79–85.

19 Marcos, A.P., Rodovalho, N.L.S., and Backes A.R. (2019). Coffee leaf rust detection using convolutional neural network. *Proceedings of 2019 XV Workshop de Visão Computacional*, pp. 1–6, São Bernardo do Campo, Brazil (9–11 September 2019). New York: IEEE.

20 Tiwari, D., Ashish, M., Gangwar, N., Sharma, A., Patel, S., and Bhardwaj, S. (2020). Potato leaf diseases detection using deep learning. *Proceedings of International*

Conference on Intelligent Computing and Control Systems, pp. 1–6, Madurai, India (13–15 May 2020). New York: IEEE.

21 Rangarajan, A.K., Purushothaman, R., and Ramesh, A. (2018). Tomato crop disease classification using pre-trained deep learning algorithm. *Procedia Computer Science* 133: 1040–1047. https://doi.org/10.1016/j.procs.2018.07.070.

22 Casadiego, Y.A.S. (2020). Inteligencia artificial (IA) y sus aplicaciones en la agricultura moderna. Artificial intelligence (AI) and its applications in modern agriculture, Colombia (September 2022), pp. 1–11.

23 Rehman, A., Iqbal, M.A., Xing, H., and Ahmed, I. (2021). COVID-19 detection empowered with machine learning and deep learning techniques: a systematic review. *Applied Sciences* 11 (8): 1–21. https://doi.org/10.3390/app11083414.

24 Rehman, A., Saba, T., Kashif, M. et al. (2022). A revisit of internet of things technologies for monitoring and control strategies in smart agriculture. *Agronomy* 12 (1): 1–21. https://doi.org/10.3390/agronomy12010127.

11

An Overview of Predictive Maintenance and Load Forecasting

Nand Kishor Gupta[1], *Vivek Upadhyaya*[1], *and Vijay Gali*[2]

[1] *Department of Electrical & Electronics Engineering, Poornima University, Jaipur, India*
[2] *Department of Electrical & Computer Engineering, Nazarbayev University, Astana, Kazakhstan*

11.1 Introduction

With the advent of predictive maintenance (PdM) and load forecasting, the old adage that "prevention is better than cure" has taken on a whole new meaning in the realm of industrial operations and the distribution of energy. The asset management, maintenance methods, and energy optimization practices of many businesses will need to undergo a paradigm shift as a result of these two interconnected areas. More and more, automation is making our lives easier. Data-driven insights are very helpful for ensuring important machines last longer, lowering downtime, and improving business efficiency.

Most maintenance practices in the past have been planned or reactive, which has led to breaks that were not expected, costly repairs, and production stops. The creation of PdM, on the contrary, has made it possible to be more proactive. The Internet of Things (IoT), data analytics, and machine learning are all used in PdM to figure out when things will break. At the same time, load forecasting has evolved as an essential component of energy management. This provides utilities with the ability to predict customer demand patterns and optimize energy generation and distribution.

This chapter will begin an in-depth investigation of the interconnected fields of PdM and load forecasting. We are going to go into the fundamental ideas, approaches, benefits, and challenges, as well as the real-world applications, that highlight the significance of these things in the contemporary industrial and

Intelligent Techniques for Predictive Data Analytics, First Edition.
Edited by Neha Singh, Shilpi Birla, Mohd Dilshad Ansari, and Neeraj Kumar Shukla.

Published 2024 by John Wiley & Sons, Inc.

energy scene. It is essential for engineers, academics, managers, and politicians alike to have a comprehensive understanding of PdM and load forecasting as automation continues to have a profound impact on the business world and the electrical systems that support it. This chapter is divided into three sections, which are outlined in Figure 11.1.

Both the manufacturing and energy industries throughout the world have seen significant benefits from implementing PdM and load forecasting. Because of their ability to foresee impending failure, maximize the effectiveness of resource allocation, and enhance the rate at which energy is distributed, the future will be more sustainable and robust thanks to their contributions. To achieve levels of operational excellence and energy sustainability that have never been seen before, it is essential to tap into the potential of PdM and load forecasting. This is the key to unlocking levels of operational excellence and energy sustainability that have never been seen before. In this chapter, the complexities of various domains are broken down, paving the way for a complete knowledge of the ramifications of those domains and encouraging creativity in an era driven by data and automation.

11.2 PdM: Revolutionizing Asset Management

The revolutionary approach to asset management known as PdM makes use of data analytics, machine learning, and the IoT to alter the traditional maintenance practices that have been used up until now. Businesses in a wide range of fields can use this cutting-edge way to make their operations more efficient, cut down on downtime, save money, and make their most valuable assets last longer. This part of the piece will talk about the ideas, benefits, technology, and real-world uses of PdM that make it such a big deal in asset management.

11.2.1 Development of Maintenance Procedures

Maintenance practices have changed over time, which has had a huge impact on how businesses run. As this journey has gone on, there has been a shift from defensive strategies to proactive ones. The way maintenance was done in the past was reactive, which means that problems with tools were fixed after they happened. This caused downtime, higher fixed costs, and problems with how things worked. This section looks at how maintenance processes have changed over time and focuses on the important steps that had to be taken before PdM became a paradigm that changed everything. In particular, the history of upkeep practices is looked at. A guess at the main steps that make up the maintenance method is shown in Figure 11.2.

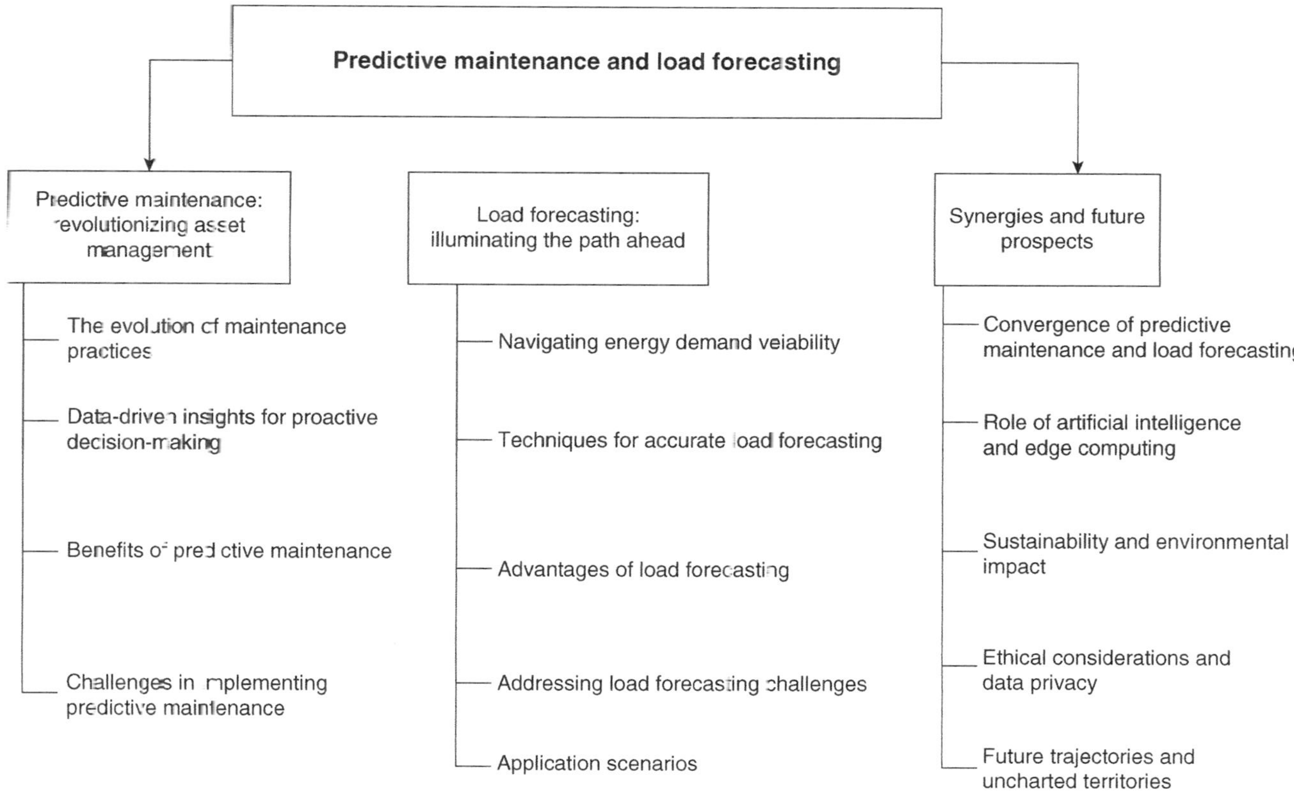

Figure 11.1 Chapter organization.

Figure 11.2 Basic stages of maintenance process.

11.2.1.1 Reactive Maintenance

This first stage was different because it used a "break-fix" approach, which meant that preventative maintenance procedures were carried out when pieces of equipment broke down. Even though it fixed problems quickly, reactive maintenance led to longer periods of downtime, less work getting done, and operations that were hard to plan for.

11.2.1.2 Preventive Maintenance

The next step meant setting up regular maintenance tasks that would be done at set times or based on usage factors. It was the goal of this plan to cut down on breakdowns by replacing or maintaining equipment before it broke down. Even though downtime was cut down a lot, preventative maintenance was not always accurate, and parts were often replaced when they were not needed.

11.2.1.3 Predictive Maintenance

With the rise of data-driven insights, advanced analytics, and the IoT, predictive repair became possible. This preventative method uses sensor data, real-time tracking, and predictive algorithms to figure out when the piece of equipment will break down. PdM looks at recurring patterns and outliers to make maintenance plans more effective. This saves money, cuts down on downtime, and makes technology last longer.

The evolution of maintenance practices from reactive to predictive represents a key transformation in industrial operations. PdM's reliance on real-time data, advanced analytics and predictive algorithms enables industries to move from a reactive to a proactive stance, thereby optimizing asset management, reducing costs and ensuring operational continuity. As technology advances, PdM is poised to play a central role in shaping the future of maintenance strategies across sectors.

The concept of multilevel anomaly detection has been introduced in prior research [1]. The initial stage involves the utilization of the Boruta algorithm [2] to process the data, thereby extracting the most significant features from the unprocessed data. The second phase involves the utilization of fire partitioning [3] in conjunction with the k-medoid partitioning algorithm [4] to ascertain the optimal number of clusters. The third phase involves employing a DBSCAN-based algorithm for the purpose of identifying and classifying potentially suspicious objects. The technique does not enforce the aggregation of all data points into a single cluster, hence resulting in infrequent identification of a distinct cluster of nonhuman data points that can be classified as abnormal. The proposed method has undergone testing on six distinct datasets sourced from online databases. The obtained findings demonstrate enhanced performance in many detection metrics, including the F1 score, true positive rate, false positive rate, and detection rate. Furthermore, with the DBSCAN algorithm, there is no requirement to explicitly specify beginning parameters as the method itself is designed to determine these values during the calculation process.

The authors presented a two-stage unsupervised approach for the detection of anomalous data using acoustic sensors [5]. Initially, the researchers partitioned the acoustic data into temporal intervals and subsequently employed three distinct techniques, namely, linear predictive coding (LPC) [6], mel frequency cepstral coefficients (MFCC) [7], and Gamma Transition (GMCC) [8], to extract features from the unprocessed data. The initial stage involves employing the BIRCH method to partition the data into many microclusters, utilizing the computed features as the basis for division. In the second stage, the merging of groups is performed based on the distance between centroids. Specifically, if the distance between two centroids is found to be below a predetermined threshold value, the merge operation is executed. Conversely, if the distance exceeds the threshold, the groups are considered to be nonmutually exclusive and are not merged. The ultimate result comprises two distinct categories: the first category encompasses action points, while the second category encompasses unusual events. In addition to the inclusion of adverse stimuli, examinations are administered to gather contextual information through the utilization of auditory cues such as gunshots, shattered glass, sirens, and vocal expressions of distress. The approach demonstrated favorable outcomes, as evidenced by measurement accuracy, memory, and F1 score surpassing 90%. The F1 score and accuracy index exhibit superior performance compared with the one-step approach that solely relies on the macro clustering method.

The authors in reference [9] introduced a decentralized architecture wherein the task of short-term anomaly detection is carried out at sensor nodes via an unsupervised artificial neural network (ANN) algorithm. During this stage, the data collected from the sensor nodes is consolidated and analyzed to enable seamless integration of the information. The subsequent phase will manifest within the cloud infrastructure and is expected to be promptly detected. Examine the association between sensors that exhibit a high degree of correlation. Hence, whereas short-term analysis entails assessing the likelihood of uncertainty and the temporal span during which data will be impacted, cloud computing necessitates a lengthier duration and relies on past data to determine sensor replacement and anomaly detection. It should be noted that the authors employ restricted Boltzmann machines (RBMs) to leverage the notion of creative games [10] and specifically to train ANN techniques in a cloud computing environment, considering the limiting capabilities of sensor points. Furthermore, it should be noted that the sensor node does not keep any data on its edge. This implies that not only the algorithm but also other relevant information is not written on the device. The authors of [11] offer a system for tracking and distributing. The solution's architecture comprises a collection of low-power devices and heads, which are energy-efficient devices positioned in proximity to the sensor nodes. The detection model was initially constructed using principal component analysis (PCA) [12, 13] in the offline training phase to compute the eigenvectors and eigenvalue matrices of the training data. At this juncture, the calculation of parameter inequality is performed for each point, utilizing the eigenvectors and eigenvalues matrix. The process of negative detection involves identifying variable values that fall beyond the established calculation boundaries during the course of the investigation. The averaging of data by the sensor points facilitates the generation of new patterns through the transmission of data to the board. The board then utilizes the data obtained from many sensor points to construct novel global models, which are subsequently transmitted back to the nodes. The efficacy of the approach was evaluated through the generation of synthetic objects and the subsequent input of data utilizing the sensor network provided by the Intel Berkeley Research Laboratory (IBRL). The findings indicate that the integration of local and global defect detection models yields superior outcomes in comparison to solely employing local models. When the approach was juxtaposed with the distribution of findings derived from time–space relationships, the outcomes indicated that just one way exhibited superior performance in terms of value analysis, while the measurement system was deemed inaccurate. However, the authors discovered that this particular method lacked comprehensiveness.

11.2.2 Data-Driven Insights for Proactive Decision-Making

Particularly in the context of PdM and optimization techniques, the era of data-driven insights has brought about a new paradigm for proactive decision-making in sectors. This section goes deeper into the revolutionary impact of data-driven insights, which are made possible by the IoT, big data analytics, and machine learning algorithms. The end goal of this section is to attain proactive and informed decision-making.

11.2.2.1 Role of IoT and Sensor Networks

The IoT has completely altered the processes of data collection and transmission. A continuous flow of data is produced as a result of the continuous collection of real-time information by sensors that are incorporated into machines, equipment, and processes. These sensor networks enable the monitoring of factors, such as temperature, pressure, vibration, and performance metrics, and they provide a comprehensive perspective of the operating circumstances.

11.2.2.2 Big Data Analytics and Machine Learning

Tools for sophisticated analytics are necessary because of the sheer volume and complexity of the data provided by IoT devices. Big data analytics tools, when combined with machine learning algorithms, sort through this data to find patterns, anomalies, and trends that might not be seen using more conventional methods. These algorithms learn from previous data and modify their behavior in response to changing situations, which results in improved forecast accuracy.

11.2.2.3 Predictive Algorithms

Data-driven insights enable the creation of predictive algorithms that identify early signs of equipment degradation or impending failures. By comparing real-time data with historical records, these algorithms recognize deviations from normal operating patterns. This proactive approach allows maintenance teams to intervene before failure occurs, minimizing downtime and optimizing resource allocation.

11.2.2.4 Real-Time Monitoring and Alerts

The integration of real-time monitoring with data analytics empowers industries to receive instant alerts when anomalies are detected. Maintenance teams can respond promptly, scheduling maintenance activities based on actual equipment conditions rather than generic time intervals.

11.2.2.5 Benefits of Data-Driven Insights

The adoption of data-driven insights for proactive decision-making offers numerous advantages. By identifying issues early, industries can avoid catastrophic failures, optimize maintenance schedules, extend equipment lifespans, and enhance operational efficiency. Moreover, predictive insights contribute to a reduction in operational costs and unplanned downtime.

11.2.2.6 Limitations and Challenges

Despite its potential, leveraging data driven insights poses challenges. Ensuring data quality, managing data integration across various systems, and training accurate machine learning models require dedicated efforts. Organizations must also address concerns related to data privacy, security, and the interpretability of complex algorithms.

With the data-driven method, learning is based on data gathered from sensors and actuators in the factory. The spread of technologies, such as IIoT and CPS [14], has made this method even more important. The gathered information can be used to look into how a part breaks down or to make a model that can guess its RUL. In recent years, researchers have looked into the data-driven approach's promise in the age of "big data." This is because it can be used with other methods in a number of different model approaches. Using statistics or machine learning, the collected data can be used to make a behavioral model or get the physical model of a certain expression, for instance. The multilayered system can handle the complexity of today's business world and use the data from older systems. This is possible because old data can be mixed with the standard for data-driven systems to make the method work better.

Some ways to handle data, such as visual search, statistics, and machine learning, can be broken down into two groups. Many models and methods for PdM have been written about, such as the Hidden Markov Model (HMM) [15], the Wiener Process Model (WPM) [16], the Gamma Process Model [17], the Proportional Hazards Model [18], and the Autoregressive Model. ARMA, which stands for equivalent linear moving average model [19]. A number of machine learning techniques are used, such as ANN [20, 21] and its variations, SVM [22], RF [23], xGBoost [24], and SOM [25]. As technologies, such as IIoT and CPS [14], have become more common, this method has become even more important. The gathered data can be used to look into how parts break down or to make behavioral models from the data and figure out their RUL. In recent years, a lot of research has been done on the promise of data-driven approaches in the age of "big data." This is mostly because these approaches can be used with other approaches in a "multi-model" way. For instance, statistical or machine learning techniques can be used on the collected data to make a model of how a system works or a physical model of a certain part. The multilayered system can handle the complexity of today's business world and use the data from older systems. This is possible because old data can be mixed with the standard for data-driven systems to make the method work better.

Some ways to handle data, such as visual search, statistics, and machine learning, can be broken down into two groups. In the research field, there are many models and methods for PdM. These include the HMM [15], the WPM [16], the Gamma Process Model [17], the Proportional Hazards Model [18], and the Autoregressive Model. ARMA, which stands for equivalent linear moving average model [19]. Several machine learning techniques are used, such as ANN [20, 21] and its variations, SVM [22], RF [23], xGBoost [24], and SOM [25]. In the next part of this chapter, we'll talk about data-driven forecasting methods and highlight the main points that each method, shown in Table 11.1, brought to the table.

Data-driven insights have reshaped decision-making from a reactive to a proactive stance, particularly in the context of PdM. The marriage of IoT, big data

Table 11.1 Summary of data-driven approaches.

	Techniques	References	Main contributions	Case study
Statistical methods	SUDM	[26]	Updates degradation models with sensor data	Simulation model
	SUDM	[27]	Advantages of SLD-based models and SUDM	Simulation model
	PF; WMQE	[28]	New health indicator	Bearing tests
	WPM; PF	[29]	RUL based on age and state	Turbo fan engines
	KS; Chi2	[30]	Automatic data labeling	Insulation of electrical machines
	WPM	[31]	Considers hard failures	Lead-acid batteries
Machine learning methods	ANN	[32]	Normal behavior model created by ANN	Wind turbine gearbox
	RBM; SOM	[33]	RBM extracts features that improve RUL prediction	Bearing data
	PF; RF	[34]	Ensemble method	Aircraft bearings; aircrafts engines
	SVM; ANN	[35]	Extended life of MEP components	MEP components
	DBF	[36]	Fusion of expert knowledge, real-time information and Configuration Parameters	Hot rolling machine
	XGBoost; RF	[37]	Applied in real environment	Consumer goods factory

analytics, and machine learning empowers industries to harness real-time information, identify patterns, and predict potential equipment failures. As technology continues to evolve, the role of data-driven insights in facilitating proactive decision-making will remain pivotal in ensuring efficient, reliable, and resilient industrial operations.

11.2.3 Benefits of PdM

PdM is a transformative strategy that harnesses data-driven insights and advanced technologies to revolutionize maintenance practices across industries. By proactively identifying potential equipment failures before they occur, it offers a myriad of benefits that reshape operational efficiency, resource utilization, and cost-effectiveness. One of its primary advantages lies in minimizing downtime through timely interventions. By analyzing real-time data and historical patterns, PdM empowers organizations to schedule maintenance during planned downtimes, reducing the impact of unexpected breakdowns and ensuring higher uptime, which directly translates to increased productivity and revenue generation.

Moreover, PdM optimizes resource allocation by providing accurate insights into the actual condition of equipment. Maintenance teams and resources are deployed based on real-time equipment health, reducing unnecessary interventions and conserving valuable resources. This not only leads to cost savings but also streamlines the overall maintenance process, making it more efficient and effective.

The strategy's ability to extend equipment lifespans is another significant benefit. By addressing minor issues before they escalate into major failures, PdM ensures that equipment operates optimally and lasts longer. This directly reduces the need for frequent replacements, cutting down on equipment acquisition costs and contributing to overall sustainability.

Cost efficiency is a prominent advantage of PdM. The strategy replaces the costly practice of reactive maintenance with planned, preventive actions. Unplanned breakdowns, emergency repairs, and excessive stockpiling of spare parts are minimized, resulting in more efficient resource utilization and reduced operational costs. This cost-effectiveness extends to overall operations, where optimized maintenance practices lead to improved profitability and financial stability.

PdM also enhances safety and reliability. By addressing potential equipment failures that might pose risks to personnel and the environment, the strategy promotes a safer working environment. Furthermore, reliable equipment operation ensures consistent product quality and customer satisfaction, enhancing the organization's reputation and market standing.

Informed decision-making is another pivotal benefit. The data-driven foundation of PdM empowers organizations to make strategic choices based on historical data and real-time information. This leads to better planning and execution, improving overall operational efficiency and resource management. Decisions are no longer reactive but are guided by insights and trends derived from extensive data analysis.

Operational efficiency is substantially improved through PdM. The reduction in downtime, optimal resource allocation, and extended equipment lifespans collectively contribute to enhanced operational efficiency. Industries can meet production targets more consistently, reduce disruptions, and maximize the output of their processes. This directly impacts overall business competitiveness and performance.

The concept of condition-based maintenance is a key innovation brought about by PdM. Unlike traditional time-based maintenance approaches, which can lead to over-maintenance or under-maintenance, PdM intervenes only when equipment shows signs of deterioration or impending failure. This precise approach eliminates the guesswork associated with scheduled maintenance, reducing unnecessary interventions and costs.

Real-time monitoring and alerts are central to the success of PdM. This instant notification system allows maintenance teams to respond promptly to anomalies detected in equipment. Early intervention prevents potential failures, mitigates operational impacts, and further contributes to reducing downtime and associated costs.

In conclusion, PdM redefines how industries approach maintenance practices. Its comprehensive benefits encompass minimized downtime, optimized resource allocation, extended equipment lifespans, cost efficiency, improved safety, enhanced operational efficiency, informed decision-making, and the strategic shift toward condition-based maintenance. By integrating data-driven insights and advanced technologies, PdM not only transforms maintenance operations but also positions organizations as industry leaders in efficiency, reliability, and cost-effectiveness.

11.2.4 Challenges in Implementing PdM

One of the biggest problems in investigating inaccuracies is that there are many causes of inaccuracies. While a fault caused by engine failure or damage contains important data, a fault caused by sensor failure, dead battery, or other external interference is considered noise and can lead to misunderstandings [38]. Therefore, a new method that uses the relationship between different sensors, reducing data errors caused by sensor problems, has been proposed [39], but only the methods mentioned in Liu et al. and ElMenshawy et al. [40, 41] clearly show the difference between noisy and abnormal data with related events. However, this method usually requires fine-tuning some parameters such as threshold and limits. Moreover, due to poor performance and new external conditions, the device needs to be modified, which requires special engineering knowledge, as shown in Liu et al. [40]. Also, when using the relationship between sensors, this uncertainty should cause significant changes in the relationship, but it will change due to the gain or other abnormality of the operation of the device. On the contrary, the tracking method proposed by Yoo [42] needs a large number of recorded data to build the model.

The absence of precise information is one of the challenges that must be overcome while building and testing a vulnerability detection system. Because of this, the neglect strategy is utilized by a number of publications [9, 43, 44]. The data that was utilized to generate performance indicators were expanded to include negative effects in addition to the results that were presented in an effort to improve this procedure. For instance, in Janjua et al. [5], the author includes sounds such as gunshots, broken glass, cries, and sirens without the typical background noise, and in Cauteruccio et al. [9], sensors are confused with light bulbs and silicone bags. Janjua et al. and Cauteruccio et al. [5, 9] both can be found online. In a real-world industrial setting, there may be instances of failure that are not immediately apparent. In addition to the noise, the collisions, and the engine damage, the actual data is unique and requires a significant amount of unique engineering work. As a consequence of this, the field of anomaly detection does not yet have such a comprehensive data set. As a result, the majority of the data sets utilized in the research given in this part incorporate synthetic mistakes into the normal behavior data sets. These flaws may not correctly represent the true behavior of the anomalous data, which is discussed further below.

The business world is becoming more and more interested in big data, and the competitive environment of the modern business world requires a swift and effective reaction. Because they may distribute the work of many agents across multiple levels, multilevel and decentralized systems have become increasingly popular in recent years. In addition, if information in the sensor network is found to be inconsistent or irrelevant, that information does not have to be saved or communicated, which allows for more efficient utilization of the available resources. It is important to note that the detection of flaws plays an important role because the results of this approach can be technically investigated in the next phase of the PdM system in two ways: valuable information may be deleted and production modifications, medical shutdowns, refueling, and so on. It is important to highlight that the detection of defects plays an important role because it is possible for the results of this approach to be technically studied in the next step of the PdM system. It is possible for sensor data to identify anomalies, which may then be incorporated into the RUL prediction model as new features. This can result in an improved level of accuracy for the prediction model.

In PdM, modeling or intelligence-based approaches are combined with data-driven ones to make complicated predictions. There are still several questions that have not been answered despite the fact that researchers have made progress in this area. A specific approach is only appropriate for use with a certain item or items; it cannot be used universally. Some data have proved beneficial in a variety of applications; for example, Lee et al. [45] showed an accuracy of better than 90% in forecasting the state of cutting tools and bearings; however, this was accomplished by using a variety of machine learning models (such as different types of SVM and ANN). In addition to that, the damage done to any other products will not be covered by this solution. In addition, modeling failures and product interactions do not make use of models such as risk factor analysis or damage trees (DFT) [46].

Bearings are one of the products that have been the subject of the most research [28, 29, 33, 47], and adequate information has been gathered through a sufficient number of experiments to simulate the product's failure. It is also simpler to produce patterns from materials for straightforward items than it is for more complicated machinery like presses or shaping machines. In addition, databases and synthetic data do not typically represent the same homogeneity that is present in real data [26, 27]. As a result, the validation of this model is not as strong as the validation of the model that was tested in an actual industrial context.

Statistical approaches, however, have a solid theoretical foundation; nonetheless, they assume that the parameters follow specific distributions, such as exponential or Weibull, among others, rather than making use of a real distribution, which might be more closely aligned with the behavior in question. In addition, the statistical approach is sometimes based on known fatigue or fracture patterns with multiple components, but the methodology cannot be repeated for other industrial equipment. This is because the known patterns are specific to the equipment in question. On the contrary, machine learning techniques have the capacity to represent highly diverse and nonlinear models, but to train the model, a substantial quantity of data and a significant amount of processing power are required.

The data-driven methodology does have its benefits; however, it does require a considerable amount of data, and failure modes are frequently ignored since there is not enough data to accurately anticipate them. In addition, this research does not combine the disturbance model with the identification of anomalies or events in any way. However, anomaly detection may also be used to detect events such as filling oil levels, altering engine settings, and changing operating circumstances. As was noted previously, anomaly detection is intended to detect faults or occurrences that can trigger urgent alerts. However, anomaly detection can also be used to detect events such as these. Anomaly detection can also be used to find conditions that can be generalized and utilized as features in predictive models, which helps to provide a common ground for PdM systems. This can be done by comparing the results of different PdM systems. Last but not least, negative detection is a technique that may be utilized to get rid of noisy data, which lowers the accuracy of model predictions. The potential of stealth detection and classification is seen in the study that was reported by Coelho et al. [48]. In this work, the authors predicted mechanical failure under high compression by combining time division and anomaly detection methods with ML classification algorithms. The F1 metric score can be increased to 22.971% when blind detection, segmentation, and machine learning are all used simultaneously. This opens the door to the possibility of increasing the usage of approaches similar to these to predict the RUL of additional assets.

11.3 Load Forecasting: Illuminating the Path Ahead

In the current energy landscape, load forecasting is an essential instrument that assists utilities and grid operators in effectively managing power output, distribution, and consumption. It gives stakeholders the ability to manage the complexity of fluctuating energy demand, maximize the use of available resources, and enthusiastically welcome the integration of renewable energy sources. This chapter delves into several facets of load forecasting, including its methodologies, advantages, and difficulties, and also several application cases.

11.3.1 Navigating Energy Demand Variability

11.3.1.1 Load Patterns and Peaks

The ability to effectively forecast load requires a fundamental understanding of load patterns as well as peak demand. The distribution of load shifts throughout the day is affected by a variety of activities, including those in the commercial and residential sectors. The stability of the grid is threatened by peaks, which frequently take place during times of extreme weather or at certain points in the calendar year.

11.3.1.2 Seasonal and Weather Influences

Seasonal and weather influences have a profound impact on electricity demand. For instance, colder winters lead to higher heating loads, while scorching summers increase air conditioning usage. Weather data integration is crucial to capture these dependencies accurately.

From a more time perspective, load forecasting can be divided into four categories. Different ML algorithms can be based on different ML algorithms: short-term load forecasting (VSTLF), short-term load forecasting (STLF), and medium-term load forecasting (STLF). STLF (MTLF) and long-term load forecast (LTLF). VSTLF is popular for load estimation in seconds to minutes. In VSTLF, future loads can be predicted based on past loads [49]. Therefore, data on temperature, economy, and land use can be selected. Rather than modeling the relationship between weather, load, time of day, and other factors affecting load, use methods that extend into the future by visualizing load patterns. Mostly automatic moving average models, negative energy sources, algorithmic genetics, etc. There are many VSTLF methods, including STLF that lasts from minutes to hours. It plays an important role in physical activity and is an important source of information for daily and weekly transactions related to production contracts and planning [50]. Since long-term transportation can be predicted by STLF, researchers often choose to develop predictive models for this area. In order to develop better short-term forecast models, the factors affecting the load need to be well understood. Some of these factors include weather, season, type of region, and time of day, etc. The important thing is integration with the demand of the load because the demand is different at every hour of the day. MTLF is often used to predict loads over days and months [51]. Winter or summer peak, etc. LTLF is popular in seasonal variable load forecasts like LTLF for all time periods from weeks to years [52]. Consider historical shipping and weather information, the number of customers in a group, specific products in an area, and more.

Among many methods, STLF is the most popular. Due to its relationship with other types of forecasting, it plays an important role in developing business strategies and improving business conditions for power systems. STLF can be converted to MTLF and LTLF by adding a variable to STLF and calculating the model over a long horizon. On the contrary, the VSTLF model can be used by STLF by adding the previous clock's load as part of the STLF model input. From the short-term load forecast, the autocorrelation between the current hour and the previous hour's load will be obtained. It is also possible to collect historical data and create new STLF-based series. A short-term forecast can be obtained by estimating future residuals and adding this to the short-term forecast. Details of the prediction are shown in Figure 11.3.

11.3.2 Techniques for Accurate Load Forecasting

11.3.2.1 Time Series Analysis

Time series analysis involves examining historical load data to identify patterns and trends. Techniques like autoregressive integrated moving average (ARIMA) [53–55] and seasonal decomposition of time series (STL) are widely used for short-term forecasting.

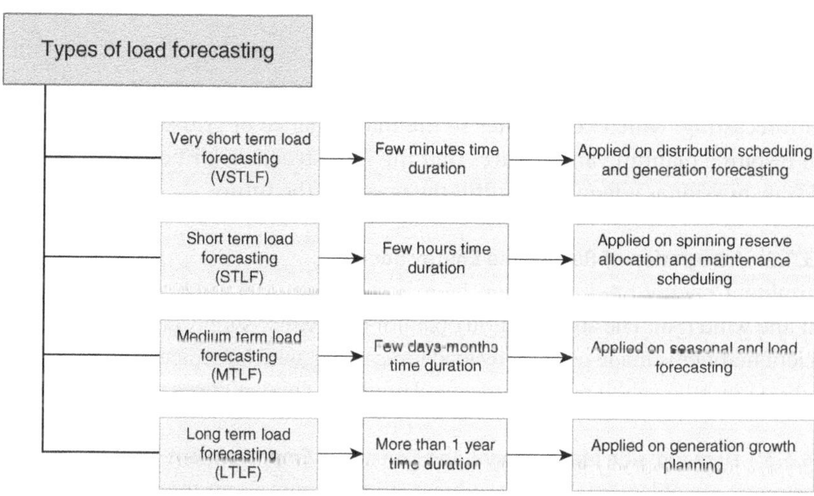

Figure 11.3 Types of load forecasting.

11.3.2.2 Machine Learning Models

Machine learning models, including ANNs, Support Vector Machines (SVMs), and Random Forests, excel at capturing complex, nonlinear relationships. They are particularly useful for medium and long-term forecasting when incorporating external variables like weather, economic indicators, and consumer behavior.

11.3.2.3 Hybrid Approaches

Hybrid approaches combine statistical and machine learning models to harness their respective strengths. For instance, combining ARIMA with ANNs can improve forecast accuracy by blending simplicity and predictive power.

A large number of literature is available on load forecasting analytical techniques. A summarized form for all possible load forecasting techniques used by researchers is given in Figure 11.4.

11.3.3 Advantages of Load Forecasting

Load forecasting (the process of predicting future electricity demand) offers numerous advantages across various sectors of the energy industry. Load forecasting is a cornerstone of modern energy management, offering innumerable benefits that contribute to grid stability, economic efficiency, environmental sustainability, and consumer empowerment. As technology continues to advance, load forecasting methods are becoming more sophisticated, further enhancing their benefits in the rapidly evolving energy landscape. These advantages are crucial for efficient energy management, resource allocation, and grid stability. Here are the key advantages of load forecasting.

11.3.3.1 Grid Stability and Resource Planning

The ability of utilities to balance supply with demand is facilitated by accurate load forecasting, which contributes to the maintenance of grid stability. It helps with resource planning and ensures that the infrastructure for power generation and transmission is adequate to fulfill the needs of the future.

11.3.3.2 Integration of Renewable Energy Sources

Intermittent energy production can be expected from renewable sources such as solar and wind [56]. The ability of grid operators to balance supply and demand in an efficient manner is made possible by load forecasting, which predicts when and how much energy renewable energy sources will produce. This facilitates their integration.

11.3.3.3 Demand-side Management and Consumer Empowerment

Consumers are able to make more educated judgments about the energy they use because of load projections. Consumers are encouraged to shift their energy use

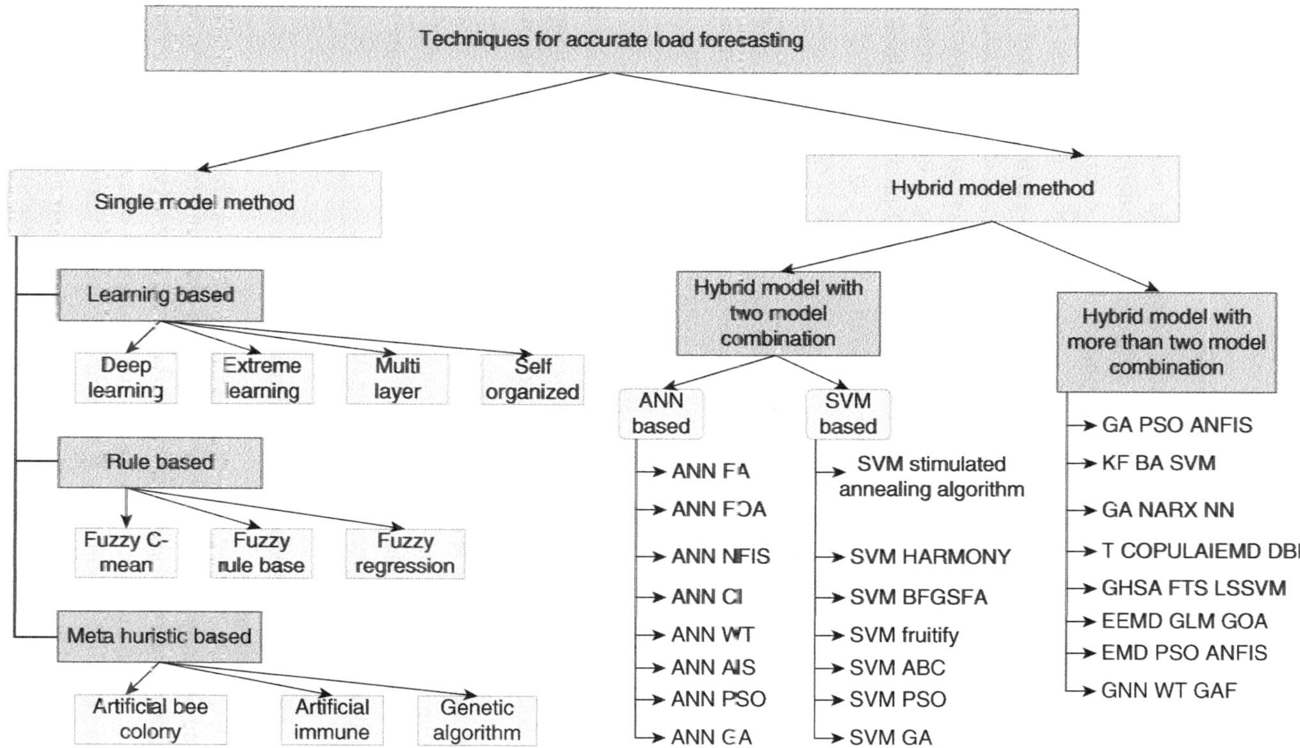

Figure 11.4 Techniques for accurate load forecasting.

to off-peak hours as part of demand-side management programs, which in turn lowers costs and total grid load.

11.3.4 Troubleshooting Load Forecasting

The process of load forecasting is an essential component of energy management despite the fact that it presents quite a few difficulties. Conquering these obstacles is absolutely necessary to achieve load estimates that are accurate and trustworthy. To effectively address these difficulties, a mix of data management best practices, advanced modeling approaches, and a dedication to staying abreast of new technological developments and trends in the sector are required. It is anticipated that the methodologies used for load forecasting will continue to advance in sophistication alongside the development of technology, thereby providing improved answers to these issues. When it comes to load predictions, these are some of the most important things to think about.

11.3.4.1 Data Granularity and Quality

When predicting load, it is very important to use proper information. Three of the issues are that the info is not always complete or correct. Things like smart monitors and other new ways to collect data are helping to solve these problems.

11.3.4.2 Incorporation of Dynamic Factors

There are a lot of changing factors that need to be taken into account when estimating load, such as new technologies, changing customer habits, and changing rules. Keeping models up to date and letting people change them is very important.

11.3.4.3 Regulatory and Market Uncertainties

There is uncertainty in load forecasts because of changes in regulations and factors in the market. Utilities need to think about these things to make sure that they are allocating resources and planning finances well.

11.3.5 Application Scenarios: From Smart Cities to Rural Electrification

Predicting how much power will be used can be useful in many situations, from making towns smarter to giving power to rural areas. In smart towns, load forecasting is very important because it helps to make the most of energy use, cut down on waste, and improve sustainability. It helps make sure that resources are spread out well in cities and makes demand response (DR) schemes possible. In rural electricity projects, load forecasting helps make off-grid energy systems that work well and do not cost too much. This makes it easier for more people to get energy, which improves the quality of life for those who do not have it. Load forecasting is a flexible tool that can be used to handle energy needs in a wide range of landscapes because it can be changed to work in both extremely developed cities and remote rural areas.

11.3.5.1 Smart Grids: Enabling DR

A big part of smart grids, which make real-time DR possible, is predicting how much power will be used. Smart cities use load forecasting to help individuals make the best use of energy, reduce waste, and make the world more sustainable.

In the context of smart grids, load forecasting is an extremely important role, particularly in the context of making DR programs possible. DR is an essential component of modern grid management because it enables utilities and consumers to modify their patterns of energy use in real time in response to shifts in both supply and demand. DR is also an important part of smart grid technology.

Real-time Data and Decision-making Load forecasting offers utilities and network operators with accurate short-term and long-term projections of the demand for electricity. Network operators are able to predict periods of high demand and potential supply limitations by continuously analyzing load projections coupled with real-time data.

Load Shifting and Peak Load Management Forecasting loads enables utilities to identify peak hours, which are often the hottest afternoons of summer and the coldest evenings of winter. Utilities can interact with users to encourage load shifting, such as operating appliances off-peak, so reducing peak loads, and optimizing overall consumption with the help of this information.

Dynamic Pricing Strategies The development of dynamic pricing models is supported by load forecasting. Real-time pricing of power can be determined by utilities with the help of load projections. Consumers are able to make educated choices based on these costs and can adapt their usage to either save money during times of high prices or consume more frequently during times of low prices.

Network Reliability and Resilience Forecasting the load on a network can assist preserve its reliability by avoiding congestion and lowering the likelihood of interruptions during times of high demand. System operators are able to guarantee that the system will continue to be stable and robust even in the face of unforeseen disasters or extreme weather conditions if they coordinate the DR activities.

Integration of Renewable Energy Smart grids frequently incorporate renewable energy sources, such as solar and wind, the production of which is subject to a degree of variability. To properly balance supply and demand, load forecasts assist determine when renewable energy will be available and how much conventional generation capacity is required.

Consumer Engagement and Empowerment Consumers are given an overview of predicted prices for electricity as well as demand patterns through the use of load forecasts. With this knowledge, customers are able to make educated decisions about their energy consumption, allowing them to tailor it to their individual preferences and financial constraints.

Incentives for Energy Efficiency When utilities can identify areas of high demand and consumption, they are able to take advantage of chances to increase energy efficiency. Accurate load predictions show these opportunities. Consumers can be motivated to lower their overall electricity usage and the energy they use by energy companies to persuade their customers to use energy efficient measures.

Peak Load Reduction DR, which is driven by load forecasting, has the potential to drastically lower peak load. As a result, utilities can save money by eliminating the need to develop extra infrastructure to manage peak demand, which results in cost savings.

To implement efficient DR programs, load forecasting is an essential component of smart grids. Load forecasting helps businesses and consumers make informed decisions, optimize energy use, save costs, boost grid dependability, and promote sustainable and efficient patterns of power consumption. It does this by giving real-time and future load projections. It is a vital building component in the transition to energy systems that are more resilient, responsive, and sustainable.

11.3.5.2 Developing Countries: Bridging the Energy Gap

In countries still in the process of building their economies, load forecasting is an extremely important component of rural electrification programs. It contributes to the construction of off-grid systems that are efficient from a financial standpoint and helps to expand access to energy, thereby enhancing the quality of life for millions of people.

11.4 Synergies and Future Prospects

A disruptive convergence of PdM and load forecasting is currently taking place in the energy business. This convergence is being driven by technological advancements as well as the increased demand for energy systems that are efficient, sustainable, and robust. In this section, the synergies and future prospects of this integration are investigated, along with the potential consequences that it may have for asset management, decision-making, sustainability, ethical considerations, and the undiscovered territories that lay ahead.

11.4.1 Convergence of PdM and Load Forecasting

The coming together of load forecasting with PdM is an exciting new breakthrough with intriguing implications for energy management and asset optimization. The improvements that have been made in data analytics, machine learning, and the IoT are causing these two historically separate sectors to become increasingly interwoven with one another. This convergence provides a number of opportunities and benefits that could lead to energy systems that are more efficient and reliable, such as the following.

11.4.1.1 Holistic Asset Management and Increasing Efficiency

End-to-end asset management is made possible by the confluence of PdM and load forecasting technologies. Energy suppliers are able to monitor the health of their infrastructure and proactively fix problems, thereby minimizing downtime and optimizing asset efficiency. This is made possible through the utilization of real-time data and predictive analytics. This synergy not only improves the reliability of the system but also lessens the amount of money spent on operating it and makes the equipment last longer.

11.4.1.2 Real-time Decision-Making for Energy Systems

The ability to make decisions in real time is essential in power systems to immediately react to shifts in demand, interruptions in the grid, and failed pieces of equipment. Energy suppliers now have the ability to make decisions based on real-time data thanks to the combination of PdM and load forecasting, which helps to maintain efficient generation, distribution, and load balancing. This capacity improves the reliability of the network and decreases the likelihood of interruptions.

11.4.2 The Role of Artificial Intelligence (AI) and Edge Computing

11.4.2.1 Edge Analytics for Localized Statistics

The combination of AI with edge computing brings data processing closer to the original source of the data. When applied to power systems, this permits the generation of localized insights that may be used for PdM and load forecasting. Edge devices that are equipped with AI algorithms are able to perform data analysis in real time, discover anomalies, and predict failures of edge devices. This results in a decreased latency and guarantees a quicker response.

11.4.2.2 Automation of Maintenance and Energy Operations Controlled by AI

AI plays a key role in the automation of maintenance processes and energy operations. AI-driven PdM algorithms can autonomously schedule repairs, order replacement parts, and even predict maintenance personnel requirements. In addition, AI-driven load forecasting can optimize energy generation and distribution, reducing waste, and emissions. This synergy improves operational efficiency while reducing human intervention.

11.4.3 Sustainability and Environmental Impact

11.4.3.1 Reduced Carbon Footprint through Optimized Operations

The integration of PdM and load forecasting contributes to sustainability by optimizing energy operations. It allows for more efficient use of resources, minimizing energy waste and carbon emissions. By preventing equipment failures and optimizing load distribution, the energy sector can significantly reduce its environmental impact.

11.4.3.2 Integration of Electric Vehicles (EVs) and Energy Storage

As EVs and energy storage systems become more prevalent, load forecasting becomes even more critical. The integration of these technologies into energy systems requires accurate predictions of when and where energy will be needed. PdM ensures the reliability of EV charging infrastructure and energy storage systems, contributing to grid stability and a cleaner energy mix.

11.4.4 Ethical Considerations and Data Privacy

11.4.4.1 Balancing Data Utilization and Consumer Privacy

The synergy between PdM and load forecasting raises ethical concerns regarding data utilization and consumer privacy. While collecting and analyzing vast amounts of data is essential for accurate predictions, it must be done responsibly. Striking a balance between data utilization and protecting consumer privacy is paramount. Regulations and policies must be in place to safeguard sensitive information.

11.4.4.2 Transparency and Accountability in Automated Systems

As AI-driven systems become more autonomous, transparency and accountability are critical. It is essential to maintain transparency in how predictions are made, and actions are taken. Furthermore, establishing accountability for system failures is crucial to build trust in these automated systems.

11.4.5 Future Trajectories and Uncharted Territories

11.4.5.1 Cognitive PdM: Merging AI and Human Expertise

The future may see the evolution of cognitive PdM systems that combine AI capabilities with human expertise. These systems would leverage AI for data analysis and predictions while involving human experts for decision validation and complex problem-solving. This fusion of AI and human knowledge can lead to more robust and adaptive systems.

11.4.5.2 Load Forecasting in a Decentralized Energy Landscape

The energy landscape is moving toward decentralization with distributed energy resources (DERs) and microgrids. Load forecasting will need to adapt to these changes, predicting demand patterns in a more complex and distributed environment. Integration with PdM in these decentralized systems will be crucial for optimizing the performance of DERs.

11.5 Conclusion

PdM redefines how industries approach maintenance practices. Its comprehensive benefits encompass minimized downtime, optimized resource allocation, extended equipment lifespans, cost efficiency, improved safety, enhanced operational efficiency, informed decision-making, and the strategic shift toward condition-based maintenance. By integrating data-driven insights and advanced technologies, PdM not only transforms maintenance operations but also positions organizations as industry leaders in efficiency, reliability, and cost-effectiveness.

Load forecasting is indispensable in modern energy management, offering a roadmap to navigate energy demand variability, optimize resource allocation, and facilitate the integration of renewable energy sources. Advances in techniques, data quality, and adaptability are propelling load forecasting into a pivotal role in shaping the future of energy sustainability and grid resilience. From smart cities to rural electrification efforts, load forecasting is illuminating the path ahead toward a more efficient and sustainable energy future.

The convergence of PdM and load forecasting represents a transformative shift in the energy industry. This synergy promises holistic asset management, real-time decision-making, sustainability, and automation of energy operations. However, it also raises ethical considerations and challenges in data privacy. As the energy sector continues to evolve, the integration of these two disciplines will play a pivotal role in shaping a more efficient, sustainable, and resilient energy future, ushering in new trajectories and uncharted territories.

References

1 Garg, S., Kaur, K., Batra, S. et al. (2020). A multi-stage anomaly detection scheme for augmenting the security in IoT- enabled applications. *Future Generation Computer Systems* 104: 105–118.

2 Kursa, M.B., Jankowski, A., and Rudnicki, W.R. (2010). Boruta – a system for feature selection. *Fundamenta Informaticae* 101 (4): 271–285.

3 Kar, A.K. (2016). Bio inspired computing – a review of algorithms and scope of applications. *Expert Systems with Applications* 59: 20–32.

4 Lee, I. and Yang, J. (2009). Common clustering algorithms. In: *Vol. 2 Comprehensive Chemometrics* (ed. S.D. Brown, R. Tauler, and B. Walczak), 577–618. Elsevier.

5 Janjua, Z.H., Vecchio, M., Antonini, M., and Antonelli, F. (2019). IRESE: an intelligent rare-event detection system using unsupervised learning on the IoT edge. *Engineering Applications of Artificial Intelligence* 84: 41–50.

6 Shaughnessy, D.O. (1988). Linear predictive coding. *IEEE Potentials* 7 (1): 29–32.

7 Bansal, P., Imam, S.A., and Bharti, R. (2015). Speaker recognition using MFCC, shifted MFCC with vector quantization and fuzzy. *2015 International Conference on Soft Computing Techniques and Implementations (ICSCTI)*, pp. 41–44, Faridabad, India (8–10 October 2015). New York: IEEE.

8 Valero, X. and Alias, F. (2012). Gammatone cepstral coefficients: biologically inspired features for non-speech audio classification. *IEEE Transactions on Multimedia* 14 (6): 1684–1689.

9 Cauteruccio, F., Fortino, G., Guerrieri, A. et al. (2019). Short-long term anomaly detection in wireless sensor networks based on machine learning and multi-parameterized edit distance. *Information Fusion* 52: 13–30.

10 Shin, H., Lee, J.K., Kim, J., and Kim, J. (2017). Continual learning with deep generative replay. In: *Advances in Neural Information Processing Systems*, vol. 2017, 2991–3000.

11 Rassam, M.A., Maarof, M.A., and Zainal, A. (2018). A distributed anomaly detection model for wireless sensor networks based on the one-class principal component classifier. *International Journal of Sensor Networks* 27 (3): 200–214.

12 Xie, Z., Quirino, T., Shyu, M.L. et al. (2006). UNPCC: A novel un-supervised classification scheme for network intrusion detection. *Proceedings – International Conference on Tools with Artificial Intelligence, ICTAI*, pp.743–750, Arlington, VA (13–15 November 2006). New York: IEEE.

13 Rassam, M.A., Zainal, A., and Maarof, M.A. (2012). One-class principal component classifier for anomaly detection in wireless sensor network. *2012 Fourth International Conference on Computational Aspects of Social Networks (CASoN)*, pp. 271–276, Sao Carlos, Brazil (21–23 November 2012). New York: IEEE.

14 Rajkumar, R., Lee, I., Sha, L., and Stankovic, J. (2010). Cyber-physical systems: the next computing revolution. *Proceedings - Design Automation Conference ACM Press*, pp. 731–736, Anaheim, CA (13–18 June 2010). New York: IEEE.

15 Rabiner, L. (1989). A tutorial on hidden Markov models and selected applications in speech recognition. *Proceedings of the IEEE* 77 (2): 257–286.

16 Iqbal, I.M. and Aziz, N. (2011). Comparison of various Wiener model identification approach in modelling nonlinear process. *2011 3rd Conference on Data Mining and Optimization (DMO)*, pp.134–140, Putrajaya, Malaysia (28–29 June 2011). New York: IEEE.

17 Singpurwalla, N. and Youngren, M. (1993). Multivariate distributions induced by dynamic environments. *Scandinavian Journal of Statistics* 20 (3): 251–261.

18 Leemis, L.M. (2010). Variate Generation in Reliability. In: *Simulation Methods for Reliability and Availability of Complex Systems*, Vol. 36 Springer Series in Reliability Engineering (ed. J. Faulin), 85–103. London: Springer.

19 Box, G.E.P., Jenkins, G.M., and Reinsel, G.C. (2008). *Time Series Analysis, Wiley Series in Probability and Statistics*. Wiley.

20 Namuduri, S., Narayanan, B.N., Davuluru, V.S.P. et al. (2020). Review—deep learning methods for sensor based predictive maintenance and future perspectives for electrochemical sensors. *Journal of the Electrochemical Society* 167 (3): 037552.

21 Schmidhuber, J. (2015). Deep learning in neural networks: an overview. *Neural Networks* 61: 85–117.

22 Hsu, C.-W. and Lin, C.-J. (2002). A comparison of methods for multiclass support vector machines. *IEEE Transactions on Neural Networks* 13 (2): 415–425.

23 Breiman, L. (2001). Random forests. *Machine Learning* 45 (1): 5–32.

24 Chen, T. and Guestrin, C. (2016). XGBoost. *Proceedings of the 22nd ACM SIGKDD International Conference on Knowledge Discovery and Data Mining*, pp. 785–794, ACM, New York.

25 Kohonen, T. (1990). The self-organizing map. *Proceedings of the IEEE* 78 (9): 1464–1480.

26 Kaiser, K.A. and Gebraeel, N.Z. (2009). Sensor-based degradation models. *IEEE Transactions on Systems, Man, and Cybernetics* 39 (4): 840–849.

27 You, M.Y., Liu, F., Wang, W., and Meng, G. (2010). Statistically planned and individually improved predictive maintenance management for continuously monitored degrading systems. *IEEE Transactions on Reliability* 59 (4): 744–753.

28 Lei, Y., Li, N., Gontarz, S. et al. (2016). A model-based method for remaining useful life prediction of machinery. *IEEE Transactions on Reliability* 65 (3): 1314–1326.

29 Li, N., Lei, Y., Yan, T. et al. (2019). A Wiener-process-model-based method for remaining useful life prediction considering unit-to-unit variability. *IEEE Transactions on Industrial Electronics* 66 (3): 2092–2101.

30 Gianoglio, C., Ragusa, E., Bruzzone, A. et al. (2020). Unsupervised monitoring system for predictive maintenance of high voltage apparatus. *Energies* 13 (5): 1109.

31 Hu, J. and Chen, P. (2020). Predictive Maintenance of Systems Subject to Hard Failure Based on Proportional Hazards Model. *Reliability Engineering & System Safety* 196: 106707.

32 Garcia, M.C., Sanz-Bobi, M.A., and del Pico, J. (2006). SIMAP: intelligent system for predictive maintenance. Application to the health condition monitoring of a windturbine gearbox. *Computers in Industry* 57 (6): 552–568.

33 Liao, L., Jin, W., and Pavel, R. (2016). Prognosability regularization for prognostics and health assessment. *IEEE Transactions on Industrial Electronics* 63 (11): 7076–7083.

34 Li, Z., Wu, D., Hu, C., and Terpenny, J. (2019). An ensemble learning-based prognostic approach with degradation-dependent weights for remaining useful life prediction. *Reliability Engineering & System Safety* 184: 110–122.

35 Cheng, J.C., Chen, W., Chen, K., and Wang, Q. (2020). Data-driven predictive maintenance planning framework for MEP components based on BIM and IoT using machine learning algorithms. *Automation in Construction* 112: 103087.

36 Ruiz-Sarmiento, J.R., Monroy, J., Moreno, F.A. et al. (2020). A predictive model for the maintenance of industrial machinery in the context of industry 4.0. *Engineering Applications of Artificial Intelligence* 87: 103289.

37 Ayvaz, S. and Alpay, K. (2021). Predictive maintenance system for production lines in manufacturing: a machine learning approach using IoT data in real-time. *Expert Systems with Applications* 173: 114598.

38 Li, D., Wang, Y., Wang, J. et al. (2020). Recent advances in sensor fault diagnosis: a review. *Sensors and Actuators, A: Physical* 309: 111990.

39 Zhao, B., Li, X., Li, J. et al. (2020). An area-context-based credibility detection for big data in IoT. *Mobile Information Systems* 2020: 1–12.

40 Liu, Y., Dillon, T., Yu, W. et al. (2020). Noise removal in the presence of significant anomalies for industrial IoT sensor data in manufacturing. *IEEE Internet of Things Journal* 7 (8): 7084–7096.

41 ElMenshawy, D. and Helmy, W. (2018). A correlation based approach to differentiate between an event and noise in internet of things. *International Journal of Advanced Computer Science and Applications* 9 (12): 79–83.

42 Yoo, Y.J. (2020). Data-driven fault detection process using correlation based clustering. *Computers in Industry* 122: 103279.

43 Gupta, R.A. and Gupta, N.K. (2015). A robust optimization based approach for microgrid operation in deregulated environment. *Energy Conversion and Management* 93: 121–131.

44 Gupta, N. (2014). Generation scheduling at PCC in grid connected microgrid. *International Conference on Recent Advances and Innovations in Engineering (ICRAIE-2014)*, pp. 1–5, Jaipur, India (9–11 May 2014). New York: IEEE.

45 Lee, W.J., Wu, H., Yun, H. et al. (2019). Predictive maintenance of machine tool systems using artificial intelligence techniques applied to machine condition data. *Procedia CIRP* 80: 506–511.

46 Dugan, J., Bavuso, S., and Boyd, M. (1992). Dynamic fault-tree models for fault-tolerant computer systems. *IEEE Transactions on Reliability* 41 (3): 363–377.

47 Lee, J., Ni, J., Djurdjanovic, D. et al. (2006). Intelligent prognostics tools and e-maintenance. *Computers in Industry* 57 (6): 476–489.

48 Coelho, D., Costa, D., Rocha, E.M. et al. (2022). Predictive maintenance on sensorized stamping presses by time series segmentation, anomaly detection, and classification algorithms. *Procedia Computer Science* 2022: 1184–1193.

49 Katzir, L. (2015). The effect of system characteristics on very-short-term load forecasting. *Przegląd Elektrotechniczny* 1 (11): 121–125.

50 Haq, M.R. and Ni, Z. (2019). A new hybrid model for short-term electricity load forecasting. *IEEE Access* 7: 125413–125423.

51 Abu-Shikhah, N., Elkarmi, F., and Aloquili, O.M. (2011). Medium-term electric load forecasting using multivariable linear and non-linear regression. *Smart Grid Renewable Energy* 2 (2): 126–135.

52 Ye, S., Zhu, G., and Xiao, Z. (2012). Long term load forecasting and recommendations for China based on support vector regression. *Energy Power Engineering* 4 (5): 380–385.

53 Gupta, N., Sharma, K.G., Mangal, A. et al. (2022). Solar power predictions in stochastics framework. *Proceedings of International Conference on Computational Intelligence and Emerging Power System: ICCIPS 2021held at Engineering College Ajmer, India*, pp. 223–233. Singapore: Springer.

54 Gupta, R.A. and Gupta, N. (2014). Robust microgrid operation considering renewable power uncertainties. *2014 Eighteenth National Power Systems Conference (NPSC)*, Guwahati, India (18–20 December 2014). New York: IEEE, pp. 1–6.

55 Gupta, N. (2016). Investigations on microgrid interconnection to distribution system. Doctoral dissertation. MNIT Jaipur.

56 Sharma, K.G., Gupta, N.K., Palwalia, D.K., and Bhadu, M. (2020). Artificial intelligence-based power quality improvement techniques in WECS. In: *Congress on Intelligent Systems* (ed. H. Sharma, M. Saraswat, S. Kumar, and J.C. Bansal), 577–587. Singapore: Springer Singapore.

12

Predictive Analytics: A Tool for Strategic Decision of Employee Turnover

SMD Azash[1], Potala Venkata Subbaiah[2], and Lucia Vilcekova[3]

[1] *Department of Law, School of Law Presidency University, Bengaluru, Karnataka, India*
[2] *Department of Marketing, GITAM School of Business, GITAM (Deemed to be University), Visakhapatnam, Andhra Pradesh, India*
[3] *Faculty of Management, Comenius University Bratislava, Bratislava, Slovakia*

12.1 Introduction

Attrition plays a significant role in the IT industry and is also affected by the rate at which businesses expand. The growth of this sector of the economy was not predetermined. There is some preparation for the availability of the necessary facilities. In our country, we build the foundation first and then launch the business. These are actualities that all major IT companies have had to face. It is fascinating to learn why people switch careers so frequently because the causes for attrition vary widely. Indeed, higher income and better benefits continue to be the primary motivation for job hopping in the modern day. However, there are many factors to think about, and it is also true that people switch careers for unique reasons. At the same time, the consistent decline cannot be blamed on employees alone. Staff members routinely assess their employer's reputation, ethics, and work environment.

It is true that Indian organizations have difficulty acquiring and retaining funding. There will always be prosperous and recessionary periods. People leave an organization too quickly if its missions are inaccessible due to a lack of focus and tried-and-true techniques in the field. Attrition is most commonly caused by negative work environments. Staff members may count on a globally minded, highly professional approach to work. They hope for an atmosphere that is welcoming, comfortable, and instructive. Bossiness is implied, and businesses prefer a carrot-and-stick approach.

Intelligent Techniques for Predictive Data Analytics, First Edition.
Edited by Neha Singh, Shilpi Birla, Mohd Dilshad Ansari, and Neeraj Kumar Shukla

Published 2024 by John Wiley & Sons, Inc.

Employees expect freedom, opportunity and great treatment from the bosses, great consolation, good encouragement, and motivation. The movement from one job to another is common. They move only when they find it difficult to adjust. Monotony is also a significant factor in attrition. Employees look forward to other jobs where there is less pressure on sales and targets. The relations at the workplace can also create discontent among employees and may be the reason for attrition.

12.2 Literature Review

Mathi and Malathi [1] have studied the issue of excessive turnover by conducting research titled "Analysing the Causes of Attrition in IT Industries - Opinion of Employee and Employer." In this study, IT experts from Chennai participated. Descriptive research was used, and a representative sample was collected for analysis. In total, 500 workers and 50 businesses were surveyed using a questionnaire with free-form responses [1, 2]. The fried man test was used to analyze the data. Employees and employers appeared to agree on organizational and family issues but disagreed on human resources and job-related issues, according to the study's findings. As a result, it is important to value employees' unique perspectives if you want to keep them around and keep your turnover low.

Kumar and Melba [3] analyzed the factors that contribute to the departure of female IT workers, including job satisfaction, retention, safety, working environment, and work–life balance. Descriptive research was used, with 523 respondents serving as the sample size. A questionnaire was used to investigate the root causes of employee turnover at Technopark Thiruvananthapuram [2, 3]. The statistical method of factor analysis was employed. Changes in job requirements, remuneration policies, and enhancements to the working environment are all examples of retention methods that should be implemented by the corporation.

The IT sector is experiencing growth and is set to revolutionize tomorrow's business practices [4]. The information technology sector's push toward the fourth industrial revolution (industry 4.0) through digital transformation has acquired a significant pace on a worldwide scale. It has been estimated that by 2020, India will have spent $94 billion on information technology. However, it has become clear that the IT industry as a whole is suffering from high attrition rates or a high turnover ratio [5, 6]. The loss of critical talent, high performers, or employees trained in the latest technology areas can be a major cause for concern due to the high cost of replacement and the time lost while recruiting new employees. However, some level of attrition is natural and even desirable because it helps organizations infuse new talent and manage margins. The authors have done exploratory research, both qualitative and quantitative, to identify and categorize the elements that influence employees' decisions to quit or remain with a company. In total, 140 participants were surveyed, interviewed, or participated in a focus group, with the majority working in the IT sector. The validity of the constructs was examined using Cronbach's alpha, one-sample T-test, group testing, and independent sample tests. The areas of pay, company culture, advancement opportunities, and work–life balance were examined for trustworthiness. It was determined, and it was also proposed, that employee turnover is lower and morale is higher in companies where these factors are prioritized.

There has been a noticeable rise in the turnover rate of professional software employees in the software industry in India, making retention a difficulty for the sector [7, 10]. This chapter's primary goals are to evaluate the effects of job loss in the software industry on both individuals and businesses [8]. In total, 100 workers from 10 different software companies were surveyed using questionnaires to acquire the primary data. Employees' inclinations to leave the company were affected by both individual and propel, as shown by the data. However, some aspects of the individual have contributed considerably to the desire to leave. Introduction as a result of global outsourcing and the staggering amount of FDI coming into China, Russia, and India, there is a huge demand for, and supply of, skilled software experts in these nations. A rising incidence of attrition is a drawback of this greater competition, especially in India.

The primary purpose of this study is to lay forth the current state of affairs and the issues that BPO workers and their employers are facing [9]. As a result, a well-thought-out and methodical approach to doing the research was developed. To get the necessary data, a survey was conducted. Employees in a subset of BPOs in New Delhi's National Capital Region constituted the study's population. The research team settled on a sample size of 1200. Since a random sample method was employed to pick the respondents, they represent a wide range of ages, sexes, marital statuses, and other demographic variables. However, the data collected is limited to entry and first-level managers, where turnover is most common. The age, education, gender, marital status, and length of service are just a few of the demographic characteristics that will be gleaned from the questionnaire. It has also collected data on the causes of employee turnover in the BPO industry, such as employees' levels of job satisfaction, salary satisfaction, organizational commitment, and inclinations to leave their current position. The data were analyzed using SPSS 11.0, a statistical program designed for the social sciences.

12.3 Need and Importance of the Study

Predictive analytics is a valuable tool for strategic decision-making in various industries, including the IT sector in India. When applied to the context of employee turnover, predictive analytics can play a crucial role in helping organizations understand, anticipate, and manage their workforce more effectively. Here is a breakdown of the need and importance of predictive analytics for managing employee turnover in the Indian IT sector:

Anticipating Turnover: Predictive analytics uses historical data and advanced algorithms to identify patterns and trends. By analyzing past employee turnover data, an organization can develop models that help predict which employees are at a higher risk of leaving. This proactive approach allows the company to take preventive measures to retain valuable talent.

Cost Reduction: Employees' turnover is expensive. It involves costs related to recruitment, training, onboarding, and lost productivity. Predictive analytics can help minimize these costs by enabling organizations to intervene before an employee decides to leave, thus reducing the need for constant recruitment and training.

Talent Retention: With predictive analytics, organizations can identify the factors contributing to employee turnover. This insight allows them to implement targeted strategies to improve employee satisfaction, engagement, and work–life balance, ultimately increasing the likelihood of retaining skilled professionals.

Data-Driven Decision-Making: Predictive analytics provides data-driven insights that guide decision-making. Instead of relying on intuition or guesswork, HR and management can make informed choices based on statistical analyses, leading to more effective strategies for retaining employees.

Customized Interventions: Predictive models can pinpoint specific factors leading to turnover, such as project dissatisfaction, inadequate career growth, or poor work–life balance. This enables organizations to design tailored interventions to address these issues for individual employees or specific teams.

Better Workforce Planning: By anticipating turnover, organizations can plan for succession, identifying potential gaps that might arise due to employee departures. This allows for smoother transitions and minimizes disruptions in project timelines.

Competitive Advantage: In the competitive IT sector in India, attracting and retaining top talent is crucial for success. Companies that effectively leverage predictive analytics to reduce turnover and enhance employee satisfaction gain a competitive edge by building a stable and high-performing workforce.

Continuous Improvement: Predictive analytics is an iterative process. By continuously collecting and analyzing data, organizations can refine their models and strategies over time, leading to more accurate predictions and better decision-making.

Employee Engagement and Satisfaction: Predictive analytics can help identify trends related to employee satisfaction and engagement. This information can guide the implementation of policies and initiatives that enhance the overall work environment and employee experience.

Ethical Considerations: While predictive analytics can provide invaluable insights, it is essential to consider ethical implications, such as ensuring employee privacy and avoiding discriminatory practices based on predictive models.

12.4 Objectives of the Study

- To apply discriminate analysis as a predictive tool for predicting employee turnover.
- To classify the employee group into stayer and leaver.

12.5 Hypothesis of the Study

H01: workplace environment not conducive is a good predictor of employee turnover.
H02: limited career growth is a good predictor of employee turnover.
H03: odd working hours are a good predictor of employee turnover.
H04: stress and work–life balance are a good predictor of employee turnover.
H05: lack of proper compensation is a good predictor of employee turnover.
H06: appraisal rating is a good predictor of employee turnover.
H07: unreasonable workload is a good predictor of employee turnover.
H08: job insecurity is a good predictor of employee turnover.
H09: team leader is a good predictor of employee turnover.

12.6 Research Method

There are around 3000 individuals living in Hyderabad who are employed by a subset of the city's IT companies. The sample size refers to the total number of things that will be randomly picked from the entire population to be used in the study. In total, 300 IT workers are chosen at random for this study. Data and information for statistical inquiries are gathered in the same ways from both primary and secondary sources. New and original information is called primary data. The study's primary data comes from questionnaires, which are filled out in closed format. The majority of the research's secondary data comes from the company itself and the IT industry's archives. The researcher has chosen to use a random sampling method.

12.7 Data Analysis Procedures and Discussion

An approach based on discriminant analysis can be used to categorize the replies into subsets according to their shared and unique features. This study used a discriminant analysis to determine whether or not nine factors known to influence employee turnover (hostile work environment, lack of opportunities for advancement, long hours without breaks, high-stress levels, low pay, negative performance reviews, excessive workload, lack of job security, and poor team leadership) could accurately predict the lever status (stayer or leaver) of employees at a sample of IT firms. The researchers first did a collinearity check in SPSS version 21 to ensure that there was no perfect linearity between the predictor variables. Table 12.1 displays the results of a multicollinearity test, which revealed that the tolerance values for discriminant analysis were above 0.1 and below 10.

Table 12.2 displays the results of a statistical analysis of the group's responses to the questions "workplace environment is not conducive," "limited career growth,"

Table 12.1 Coefficients of multi-collinearity assessment.

	Unstandardized coefficients				**Collinearity statistics**	
Model	***B***	**Standard error**	***t***	**Significance**	**Tolerance**	**Variance inflation factor**
(Constant)	0.043	0.048	0.882	0.378		
Workplace environment is not conducive	0.903	0.112	8.081	0.000	0.040	4.767
Limited career growth	−0.009	0.162	−0.053	0.958	0.019	5.161
Odd working hours	0.003	0.039	0.085	0.932	0.042	3.697
Stress and work-life balance	0.002	0.086	0.022	0.982	0.068	4.685
Lack of proper compensation	0.009	0.059	0.158	0.875	0.145	6.880
Appraisal rating	−0.007	0.113	−0.063	0.950	0.039	5.494
Unreasonable workload	−0.006	0.032	−0.181	0.857	0.060	6.780
Job insecurity	0.016	0.033	0.468	0.640	0.056	7.841
Team leader	0.009	0.058	0.158	0.875	0.148	6.753

Table 12.2 Group statistics of mean and standard deviation on employee turnover (leaver status).

Group statistics					
				Valid *N* (listwise)	
Leaver status		**Mean**	**Standard deviation**	**Unweighted**	**Weighted**
Stayer	Workplace environment is not conducive	1.0667	0.25014	180	180.000
	Limited career growth	1.0944	0.29326	180	180.000
	Odd working hours	3.7333	1.45548	180	180.000
	Stress and work–life balance	1.1056	0.30813	180	180.000
	Lack of proper compensation	1.1333	0.34088	180	180.000
	Appraisal rating	1.1111	0.31515	180	180.000
	Unreasonable workload	3.7056	1.46351	180	180.000
	Job insecurity	3.6722	1.46783	180	180.000
	Team leader	1.1667	0.37372	180	180.000
Leaver	Workplace environment is not conducive	2.0000	0.00000	120	120.000
	Limited career growth	2.0000	0.00000	120	120.000
	Odd working hours	3.9250	1.31035	120	120.000
	Stress and work–life balance	2.0000	0.00000	120	120.000
	Lack of proper compensation	2.0000	0.00000	120	120.000
	Appraisal rating	2.0000	0.00000	120	120.000
	Unreasonable workload	3.9000	1.31826	120	120.000
	Job insecurity	3.9500	1.30191	120	120.000
	Team leader	2.0000	0.00000	120	120.000
Total	Workplace environment is not conducive	1.4400	0.49722	300	300.000
	Limited career growth	1.4567	0.49895	300	300.000
	Odd working hours	3.8100	1.40016	300	300.000
	Stress and work–life balance	1.4633	0.49949	300	300.000
	Lack of proper compensation	1.4800	0.50043	300	300.000
	Appraisal rating	1.4667	0.49972	300	300.000
	Unreasonable workload	3.7833	1.40819	300	300.000
	Job insecurity	3.7833	1.40819	300	300.000
	Team leader	1.5000	0.50084	300	300.000

"odd working hours," "stress and work life balance," "lack of proper compensation," "appraisal rating," "unreasonable workload," "job insecurity," and "team leader." Management role, work load, role conflict, and role ambiguity are only few of the nine independent variables whose mean values reveal a distinction between employees who stay and those who leave the organization. To give just one example, the mean values for "odd working hours," "unreasonable workload," and "job insecurity" were all lower for those in the "stayer" category than they were for those in the "leaver" category, indicating discrimination.

The study's findings corroborate those of Essiam et al. [11], who found that the effect size of factors, such as an unfriendly work environment, few opportunities for advancement, long hours, excessive stress, low pay, poor-performance reviews, an excessive workload, uncertain employment prospects, and a weak team leader in explaining employee turnover (leaver status) was practically small (Table 12.3).

Table 12.4 shows that the Wilks' Lambda value for IT workers is 0.150, which is close to the 0.0 value that indicates a significant difference between employee turnover caused by factors, such as an unwelcoming work environment, few opportunities for advancement, excessive stress, long hours, low pay, poor-performance reviews, an excessive workload, uncertain employment prospects, and a lack of a strong team leader.

Table 12.5 summarizes canonical discriminant functions, indicating that the model accounts for 66.02% of the variance in employee turnover due to nine job-related factors: unfavorable work environment, limited opportunities for advancement, unusual

Table 12.3 Box's test of equality of covariance matrices.

Box's M		107.801
F	Approximately	4.906
	Degree of freedom 1	21
	Degree of freedom 2	69 613.243
	Significance values	0.000

Tests null hypothesis of equal population covariance matrices.

Table 12.4 Wilks' lambda.

Wilks' lambda				
Test of function(s)	**Wilks' lambda**	**Chi-square**	**Degree of freedom**	**Significance**
1	0.150	556.619	9	0.000

Table 12.5 Summary of canonical discriminant functions.

Eigenvalues				
Function	**Eigenvalue**	**Percentage of variance**	**Cumulative percentage**	**Canonical correlation**
1	5.662[a]	100.0	100.0	0.922

[a] First 1 canonical discriminant functions were used in the analysis.

work hours, high-stress levels, inadequate pay, poor-performance reviews, excessive workload, lack of job security, and a poor team leader.

Standardized function coefficients and structure matrix (Tables 12.6 and 12.7) reveal that discriminatory turnover among IT workers is caused by factors such as a hostile work environment, a lack of opportunities for advancement, long hours, excessive stress, a lack of financial reward, an unfair workload, an unreliable manager, and an uncertain future. The workplace environment is not conducive, lack of career growth opportunities, stress and work–life balance, appraisal rating, inadequate compensation, and a poor team leader were the next most important predictors of employee turnover among IT workers, in that order. These aspects of workers' jobs are significant predictors of employee turnover.

Table 12.7 details the proportion of patients originally assigned to the stayer or leaver categories. Overall, 96% of cross-validated grouped cases were correctly

Table 12.6 Standardized canonical discriminant function coefficients.

	Function
	1
Workplace environment is not conducive	**0.997**
Limited career growth	−0.011
Odd working hours	0.026
Stress and work–life balance	0.003
Lack of proper compensation	0.014
Appraisal rating	−0.010
Unreasonable workload	0.047
Job insecurity	0.125
Team leader	0.015

Highest value which is near to one.

Table 12.7 Structure matrix.

Structure matrix	
	Function
	1
Workplace environment is not conducive	**0.994**
Limited career growth	**0.823**
Stress and work–life balance	**0.774**
Appraisal rating	**0.752**
Lack of proper compensation	**0.678**
Team leader	**0.594**
Job insecurity	0.041
Unreasonable workload	0.029
Odd working hours	0.028

Pooled within-groups correlations between discriminating variables and standardized canonical discriminant functions.
Variables ordered by absolute size of correlation within function.
Highest values which are having more than 0.6.

categorized based on the model's predictions of whether or not IT employees wanted to remain with the organization in the near future. The data shows that workers who plan to leave their jobs soon were identified more accurately (96%) than those who intend to remain in their current positions (100%) (Table 12.8).

12.8 Recommendations

Data Collection and Quality: Ensure that accurate and relevant data is collected, including employee demographics, performance metrics, engagement surveys, and turnover history. High-quality data is essential for building reliable predictive models.

Collaboration: Establish cross-functional teams involving HR, data analysts, and business leaders. Collaboration fosters a comprehensive understanding of turnover drivers and encourages diverse perspectives in designing retention strategies.

Feature Selection: Identify the most relevant variables affecting turnover, such as job satisfaction, project engagement, career growth opportunities, and compensation. Focus on these factors to develop more accurate predictive models.

Table 12.8 Classification results.

Classification results[a,b]					
		Leaver status	Predicted group membership: Stayer	Predicted group membership: Leaver	Total
Original	Count	Stayer	168	12	180
		Leaver	0	120	120
	Percentage	Stayer	93.3	6.7	100.0
		Leaver	0.0	100.0	100.0
Cross-validated[c]	Count	Stayer	168	12	180
		Leaver	0	120	120
	Percentage	Stayer	93.3	6.7	100.0
		Leaver	0.0	100.0	100.0

[a] 96.0% of original grouped cases correctly classified.
[b] 96.0% of cross-validated grouped cases correctly classified.
[c] Cross-validation is done only for those cases in the analysis. In cross-validation, each case is classified by the functions derived from all cases other than that case.

Model Development: Employ advanced analytics techniques, such as machine learning algorithms, to develop predictive models. Regularly update and refine these models as new data becomes available.

Ethical Considerations: Prioritize employee privacy and fairness. Ensure that predictive models do not discriminate against any group and that interventions are aimed at improving overall employee experience.

Actionable Insights: Translate predictive insights into actionable strategies. Tailor interventions to address specific concerns identified by the models, such as providing training for skill development or enhancing work–life balance.

Continuous Monitoring: Establish a feedback loop to monitor the effectiveness of implemented strategies. Regularly assess whether turnover rates are aligning with predictions and make necessary adjustments.

12.9 Conclusion

In the dynamic landscape of the Indian IT sector, managing employee turnover is a critical challenge. Predictive analytics emerges as a potent tool that can guide strategic decision-making to tackle this issue. By harnessing historical data, advanced algorithms, and cross-functional collaboration, organizations can anticipate turnover patterns and take proactive measures.

Predictive analytics not only aids in cost reduction and talent retention but also enables a shift toward data-driven HR practices. The insights gained empower organizations to tailor interventions, optimize workforce planning, and improve employee engagement. Moreover, predictive analytics provides a competitive edge by enabling companies to build a stable, skilled, and motivated workforce.

However, it is crucial to navigate ethical considerations and ensure that predictive models do not compromise employee privacy or lead to discriminatory practices. Regularly refining models and strategies based on ongoing feedback is essential for maintaining the relevance and accuracy of predictive analytics in managing employee turnover.

In the context of the Indian IT sector, where competition for top talent is intense, the use of predictive analytics for strategic decision-making regarding employee turnover can be a game-changer. Organizations that effectively implement these recommendations stand to create a work environment that attracts, retains, and nurtures the best talent, contributing to their long-term success in the industry.

References

1 Mathi, K.M. and Malathi, G. (2016). Analysing the causes of attrition in IT industries – opinion of employee and employer. *Asia Pacific Journal of Research* 1 (4): 1–20.

2 Gupta, V. (2013). An analysis of attrition: retention strategy for IT/BPO industry. *International Journal of Advance Research in Computer Science and Management Studies* 7: 434–442.

3 Kumar, R.M. and Melba, A.A. (2017). A study on women employee attrition in IT industry with special reference to Technopark, Thiruvananthapuram. *IEEE International Conference on Acoustics, Speech, and Signal Processing (ICASSP) 2017*, New Orleans, LA (5–9 March 2017). New York: IEEE.

4 Tandon, D., Gupta, S., and Tandon, N. (2020). Attrition in the information technology (IT) industry using HR analytics. *Test Engineering and Management* 83: 9880–9893.

5 Dhal, S.K. and Nayak, A.C. (2013). A study on employee attrition in BPO industries in India. *International Journal of Science and Research (IJSR)* 4: 2246–2249.

6 Loganathan, M. and Ashwini, S. (2017). A study on employee attrition and retention in manufacturing industries. *International Journal of Commerce and Management Research* 3: 43–45.

7 Thirulogasundaram, V. and Kumar, S. (2012). Assessment of individual and propel intention for job attrition on software industry: Voice from software employees in Bangalore city, India. *European Journal of Business and Management* 4 (6): 48–54.

8 James, M.J. and Faisal, U. (2013). Empirical study on addressing high employee attrition in BPO industry focusing on employee salary and other factors in Karnataka and Kerala states of India. *Research Journal of Management Sciences* 2 (9): 7–11.

9 Priya, R., Vinothini, G., and Jesu, C.D.C. (2019). A study on reasons for employee attrition in Indian BPO industry. *Journal of Advanced Research in Dynamical and Control Systems* 11: https://doi.org/10.5373/JARDCS/V11/20192752.

10 Suresh, S., Visvalingam, D., Lu, A., and Wright, B. (2020). Evaluating and improving attrition models for the retail banking industry. *2020 Systems and Information Engineering Design Symposium, SIEDS 2020,* Charlottesville, VA (24 April 2020). New York: IEEE.

11 Essiam, J.O., Mensah, M.E., Kudu, L.K., and Gyamfi, G.D. (2015). Influence of job stress on job satisfaction among university staff: Analytical evidence from a public university in Ghana. *International Journal of Economics, Commerce, and Management* 3: 1–15.

Index

Intelligent Techniques for Predictive Data Analytics, First Edition.
Edited by Neha Singh, Shilpi Birla, Mohd Dilshad Ansari, and Neeraj Kumar Shukla.

Published 2024 by John Wiley & Sons, Inc.

Printed and bound by CPI Group (UK) Ltd, Croydon, CR0 4YY

17/09/2026

14964311-0001